Graduate Study for the Twenty-First Century

Graduate Study for the Twenty-First Century

How to Build an Academic Career in the Humanities

Second Edition

Gregory M. Colón Semenza

With a Foreword by

Michael Bérubé

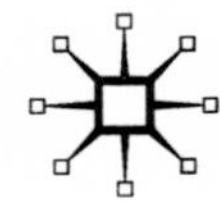

GRADUATE STUDY FOR THE TWENTY-FIRST CENTURY

First edition published in 2005 by PALGRAVE MACMILLAN® in the United States—a division of St. Martin's Press LLC, 175 Fifth Avenue, New York, NY 10010.

Where this book is distributed in the UK, Europe and the rest of the world, this is by Palgrave Macmillan, a division of Macmillan Publishers Limited, registered in England, company number 785998, of Houndmills, Basingstoke, Hampshire RG21 6XS.

Palgrave Macmillan is the global academic imprint of the above companies and has companies and representatives throughout the world.

Palgrave® and Macmillan® are registered trademarks in the United States, the United Kingdom, Europe and other countries.

ISBN: 978–0–230–10033–6

Library of Congress Cataloging-in-Publication Data

Semenza, Gregory M. M. Colón , 1972–
Graduate study for the twenty-first century : how to build an academic career in the humanities / Gregory Colon Semenza.—Rev. and updated 2nd ed.
p. cm.
ISBN 978–0–230–10033–6 (pbk.)
1. Universities and colleges—United States—Graduate work.
2. Humanities—Study and teaching (Graduate)—United States. I. Title.

LB2371.4.S46 2010
378.1'55—dc22 2009047064

A catalogue record of the book is available from the British Library.

Design by Newgen Imaging Systems (P) Ltd., Chennai, India.

Second edition: March 2010

10 9 8 7 6 5 4 3 2 1

Printed in the United States of America.

Transferred to Digital Printing in 2011

For Jane Cable and Patricia Chapman,
the two teachers . . .

and

Hans Turley (1956–2008)

Contents

Preface to the Second Edition

When I learned last year that *Graduate Study for the Twenty-First Century* has been used in some engineering, physical education, and clinical psychology graduate programs—hardly the humanities disciplines it was intended for and, to be honest, fields about which I know nothing—I realized that its success, alas, has had as much to do with the niche it fills than with any particularly mind-blowing advice it offers. That niche can still be defined today as I defined it in 2005: as an absence of field-specific practical information for would-be academics. For this reason, I am even more immensely grateful for the touchingly warm and generous responses that readers have shared with me over the past four years. The additions and corrections herein are almost entirely the result of feedback—some critical—I have received through e-mail, phone calls, and Q&As following guest lectures and workshops. The additions include a new Introduction focused on advising issues, significantly expanded material on personal voice in "The Seminar Paper" chapter, and an Afterword about the current recession and its impact on the humanities.

August 2009

Acknowledgments to the Second Edition

Thanks both to my original Palgrave editor, Farideh Koohi-Kamali, and my current one, Brigitte Shull, for immediately understanding the value of a second edition. Lee Norton at Palgrave has also been tremendously helpful.

Some people who deserved mention in the first edition and some others who deserve it now: Marco Abel, Tommy Anderson, Doreen Bell, Richard Bleiler, Reinhold Bubser, Jane Cable, Patricia Chapman, Judy B. Cheatham, Richard Cunningham, Lara Dodds, Don-John Dugas, Katherine Eggert, Molly Ferguson, Elizabeth A. Flynn, Samuel Gladden, J. Glenn Grayson, Wendy Hoofnagel, Ken Jackson, Matt Kinservik, Jon Kotchian, Paula Krebs, Ronald Levao, Charles Mahoney, Geoffrey Meigs, Ryan Netzley, Penelope Pelizzon, Cathy Schlund-Vials, Susie Schwieger, Benjamin Colón Semenza, Susan Solomon, Garrett Sullivan, Mary Udal, Dr. Virago, and Robin Worley.

Preface

I was determined, while still a graduate student, that I would someday write this book, but I had no plan to do it so early on in my professorial career. To be honest, I envisioned myself turning to it near the end of a long career; as a wiser, older man, I could reflect on decades of experience and write *the* book that would answer all of the questions worth asking about graduate school. Alas, like all young and foolish men, I've been more impetuous. Having received my Ph.D. at Penn State in 2001 with a specialization in Renaissance Literature, I've been happily employed for the past four years as an Assistant Professor of English at the University of Connecticut. In my capacity at UConn as a graduate faculty member and, more recently, Director of Graduate Studies, I've taken advantage of numerous opportunities to test many of the materials and ideas I had always hoped to include in this graduate school book. After publishing my first book in 2003, one of my colleagues asked why I didn't simply go ahead and finish what we lovingly referred to as "the grad school thing"; rather than regarding my relative youth as a liability, she explained, I should emphasize the point that graduate students need advice from those professors whose experiences have been closest to their own. Since one of the arguments of my book is that graduate school at the turn of the twenty-first century is very different than what it was 20 or even 10 years ago, she had little trouble convincing me to go ahead with it.

So why do we need another book about graduate school? Generally speaking, *Graduate Study for the Twenty-First Century* faces direct competition from four previous studies. For almost 20 years, the best book on the market has been *Getting What You Came For: The Smart Student's Guide to Earning a Master's or Ph.D.* (Farrar, Straus, and Giroux, 1992). Authored by a biology Ph.D., the book's greatest virtue is its comprehensiveness, but at the same time, its attempt to cover every aspect of the graduate experience (110 pages on "getting in" and obtaining financial aid, chapters on the historical development of MA and Ph.D. programs, etc.) means that specific matters such as lesson planning, conferencing, and publishing are treated in insufficient detail.

Major activities relevant to humanities students, such as seminar paper writing and departmental service, are simply ignored. While *Getting What You Came For* is the only book to address specific problems faced by minority students, it limits its discussion to racism, failing to address the single biggest obstacle faced by minority students, which is the burden of unfair departmental and university service. *The Ultimate Grad School Survival Guide* (Peterson's, 1996) offers sound advice about *most* of the important issues (again ignoring several major subjects such as departmental service and academic writing), but, as mentioned earlier, its excessive focus on the application process and its "soundbite" approach results in overly brief discussions of the serious issues facing today's graduate students: for example, MA and Ph.D. exams are treated in three paragraphs; conferencing, book reviewing, article publishing, and book publishing are all treated in one extremely short section. *The Grad School Handbook* (Perigee, 1998) dedicates 180 of its 232 pages to "getting in," and the remaining 50 seek only to *describe* a typical graduate program. The book fails to offer advice on such basic topics as course-work, exams, presentations, professional development, or the job search. Finally, *Playing the Game: The Streetsmart Guide to Graduate Study* (iUniverse, 2003), approaches graduate study from what can only be regarded as a comedic perspective. Authors "Frank" and "Stein" attempt to demystify the graduate experience by cracking jokes about it, which isn't necessarily a bad idea. Unfortunately, the tone is so cynical and flippant and the substance so thin that most graduate students—who must feel that the academic life is more appealing than the book's authors do—will find relatively little in *The Streetsmart Guide* that can be taken seriously.

Five characteristics distinguish *Graduate Study for the Twenty-First Century*. First, this is a book designed solely for graduate students who wish to become professors on the tenure track; it does not spend time on alternative career paths for terminal MAs or Ph.D.s. Second, the unique focus on building a professorial career means that this book dedicates a significant amount of attention to professional development issues, including publishing, attending conferences, and job searching. In a straightforward and non-condescending manner, it emphasizes how a smart and informed "streamlining" approach to graduate study and teaching can lead to both a meaningful (and relatively short) graduate career and the sort of professional accomplishments that will make you a standout on the job market. Third, *Graduate Study for the Twenty-First Century* is the only guide that recognizes the specific needs of students in the humanities. It does not assume that the concerns of a history student (or professor) are the same as those of an individual specializing in chemistry or engineering. Fourth, this book deliberately counters the tendency of the aforementioned guides to present an image of graduate school as unrelated to and unaffected by the brutal realities of late-twentieth-century and twenty-first-century politics and corporate economics. One gets the impression from many previous graduate school guides that academe is no different today from what it was 50 or 75 years ago. Finally, this book operates at a level of detail simply not found in any of the aforementioned works. Focusing in depth on such important practical matters as selecting the right seminars, making the most of exams, and constructing effective CVs, teaching portfolios, and job applications, the emphasis of this book is very much on *how to* succeed in graduate school.

2005

Acknowledgments

Usually persons fortunate enough to write this type of book base their right to do so on the amount of time they've served in academe; surely you've seen the book jackets I'm talking about: "Professor Genius has served as Director of Graduate Studies at Ivy-Bedecked University for 71 years." Especially in light of such a fact, I would like to thank Farideh Koohi-Kamali, my editor at Palgrave Macmillan, who listened so carefully when I suggested to her that today's graduate students might have something to learn from a more recent survivor of both graduate school and the humanities job market. I should also like to thank Melissa Nosal at Palgrave, who urged me to contact Farideh in the first place, and Mr. Maran Elancheran for his work overseeing the production of the book.

Three individuals in particular have been immensely supportive of this project, agreeing to read the entire manuscript and sharing their honest feedback along the way. I am grateful to Robert Hasenfratz and Jerry Phillips, true friends and dedicated members of the UConn summer writer's group, and Kathryn Hume, author of a remarkable book on the academic job market (read it), and one of the most generous individuals I know.

I am especially honored by Michael Bérubé's willingness to contribute a foreword for this book. Professor Bérubé has been, from the beginning, a tireless advocate of graduate students and an honest and ethical voice in the whirlwind.

Other individuals who warrant special mention: Sean Grass, a wonderful friend, colleague, and the bottomless source of inappropriate humor; my department head, Bob Tilton, who has made life good at UConn; Liz Jenkins, whose career has been dedicated to helping graduate students; and Jack Selzer, who taught me something about how a graduate program should be run.

To others who offered materials, support, and inspiration along the way: Ray Anselment, Doreen Bell, Richard Bleiler, Brontë Berger, Patrick Cheney, Josh Eyler, Guiseppina Iacono, Laura Knoppers, Rose Kovarovics, Ana María Gómez Laguna, Niamh O'Leary, Karen Renner,

Dave Rice, Matt Semenza, Garret Sullivan, Polya Tocheva, "Toonce," Hans Turley, Mary Udal, Reginald Wilburn, Linda Woodbridge, and the participants in my English 497 workshop, who were *forced* to read this book while it was still a work in progress.

Finally, to my truest friend and my wife, Cristina, for her unconditional support, her beautiful mind, and for Alexander—four weeks old today, and the most unbelievable joy I've ever known.

Storrs, CT
September 5, 2004

A Note on Terminology

In order to keep the advice I offer here as immediate and personal as possible, I use the second person familiar pronoun far more often than I would ever allow my undergraduate students to do. In order to prevent awkwardness in the prose, I alternate male and female pronouns by chapter; for example, whereas chapter 1 uses the terms "she" and "her," chapter 2 uses "he," "his," and "him," and so forth.

Foreword

At some point in the early 1990s, a handful of my assistant-professor colleagues at the University of Illinois at Urbana-Champaign and elsewhere decided that what the profession needed was a handbook on How to Be a Graduate Student. Not another guide on applying to graduate school, but a wholly new genre, a guide to *being in* graduate school (filled, of course, with advice on getting out of graduate school as well). Our dissertation defenses were only a few years behind us, and we had that new-recruit reformer's zeal: we knew, in fresh retrospect, what had and hadn't worked in our own graduate school careers, and now that we had assumed the responsibility of teaching and training graduate students of our own, we could see how the system rewarded the students who already knew (more or less) what they were doing with their programs of study, and how it flummoxed the students who weren't quite sure what they were doing, or who weren't quite sure how to go about doing it better.

Our discussions of How to Be a Graduate Student didn't take the form of wishing for "better" graduate students or "better prepared" graduate students. Although I was—and still am—stupefied by the phenomenon of graduate students who sit in seminars and never say a word, at the time we were primarily concerned with creating better graduate *programs*. When I arrived at Illinois, for instance, I quickly learned that the English Department had no placement director for new Ph.D.s; there was a director of graduate studies, to be sure, but no one who oversaw and guided the students who were actually looking for jobs. Instead, I found a cohort of graduate students who had been advised—why and by whom, I never learned—that their letters of application to English Department search committees should not exceed one page. "But, but, but," I stammered in surprise, "that gives you only a few sentences in which to describe your dissertation and your teaching record. Or do you save the synopsis of your research for the dissertation abstract?" The students told me that they weren't sending out dissertation abstracts, either. *Holy hamstring, Batman*, I thought—these students weren't giving themselves any chance (more precisely, they had been told not to give themselves any chance) to describe the research they'd been working on for 2 or 3 years, and they weren't giving search committees any sense of what their dissertation was arguing, and how, and why. So I went directly to the department head, full of new-recruit reformer's zeal, and volunteered for the position of placement director. A few years later, I teamed up with Cary Nelson to advocate improvements in graduate education and to support nationwide efforts to unionize graduate students. But I never got around to writing anything about How to Be a Graduate Student.

And now I don't have to, because Gregory Colón Semenza has written the ideal book on the subject. By "ideal" I mean simply this: it is sane, circumspect, and sagacious. I also mean to suggest that its sanity and circumspection are every bit as valuable as its sagacity. Semenza knows that no two humanities departments are alike, and that there is almost as much variation among graduate programs as there is among graduate students. He remembers well how terrifying it is to face your first class as a teacher, and he knows how difficult it is to try to explain to your parents—or your loved one's parents—what you're doing (and hoping to do) with your life. He knows what it's like to balance the demands of profession and family, and he knows what it's like to mediate among differently-minded members of a comprehensive-exam committee. Best of all, he knows how the academic professions really work, right down to the invisible but critical minutiae of departmental committee service and the tricky question of when it's all right to ask a journal editor what happened to the essay you submitted last spring. The result is that *Graduate Study for the Twenty-First Century* might just be the least idiosyncratic—that is, the most reliable—book I have ever read about academe and its inhabitants.

If you're thinking about joining academe and its inhabitants, I simply cannot press this point strongly enough—because if there's one thing that makes career advice worthless (or worse), whether you're a prospective graduate student, a harried ABD, or a new assistant professor, it's idiosyncrasy. And academe, being academe, is full of it. I recall vividly the closing moments of one dissertation defense in which a committee member, addressing the question of how the candidate could best revise her work for publication, turned to the rest of the committee and said, "about how much of the dissertation, would you say, should wind up in the finished manuscript?" Before I could reply, "well, it all depends on the dissertation, and this one's quite strong," he revealed that the question was not really a question, as he graciously answered it himself: "that's right, about forty percent." (To this day I savor the "that's right.") Fortunately, I happened to be the director of that committee, and could advise the student later that evening, "ix-nay on the orty-fay ercent-pay—you simply need to tighten the last two chapters and write a new conclusion. Come talk to me next week." Or I might mention the colleague who advised a student not to submit an essay to the journal *Cultural Critique* because he'd never heard of it. Or the colleague who advised her students to request letters of recommendation from full professors, and only full professors. In each of these cases, students got terrible career advice, and the only reason I know about this terrible advice is that the students came to me and asked, "is that right?" To which, of course, the short answer is no—and the longer answers can be found in these very pages.

That's not to say that Greg Semenza hasn't established his own distinctive and salient voice in the course of writing this book. On the contrary, from start to finish, *Graduate Study in the Twenty-First Century* reads as if it's written by a trusted friend and mentor—someone stern enough to tell you that if you're not going to read a Victorian novel on your own you shouldn't be in graduate school; someone patient enough to walk you through the process of submitting proposals and drafting papers for conferences; someone sympathetic enough to let you in on what I call the "first pancake phenomenon," namely, the fact that it's nearly impossible to get a course "down" the first time you teach it. Moreover, Semenza has done well to have framed this book as what he calls a " 'working class' approach to graduate study," since no matter where you're thinking of applying, dear reader, no matter where you may be studying now, the vast majority of jobs in the academic profession are to be found neither at Yale nor at Oberlin. Recognizing this fact of life is crucial for anyone who aspires to a career in academe—as is realizing that one can have a perfectly satisfying, stimulating academic career elsewhere than at Yale or Oberlin.

Indeed, one of the most remarkable features of this book is its clear-sightedness about the actual state of the academic job scene. Semenza does not blink away the legions of adjuncts, part-timers, and day-laborers who toil in the groves of academe; on the contrary, he's woven into his discussion of the academic profession a bright thread of warning about the degree to which academic jobs themselves have been deprofessionalized. This feature of academe is sometimes all too obscure to long-tenured faculty, some of whom have lost touch not merely with the realities of graduate education but with the working conditions of almost half of the academic workforce. In this respect as in many others, Semenza is quite right to remind us that (as one of his colleagues put it, in the course of encouraging him to write this book) "graduate students need advice from professors whose experiences have been closest to their own." It's not that most of us older folk, after 10 or 20 years, begin to lose our reformer's zeal; some of us never do. But as people like me enter their mid-forties and their mid-careers, they inevitably lose even the memory of the sense of what it's like to get that first article acceptance, what it's like to present that first conference paper, what it's like to send out those first couple dozen letters to search committees knowing full well that less than half of new Ph.D.s in your field will wind up with tenure-track jobs. (Indeed, very few people in my own Ph.D.-candidate cohort in the late 1980s had presented papers at *any* conferences, and only a tiny handful of us had published essays before entering the job market. Already that period, recent as it is, looks distant and sepia-tinged.) Tenured professors like me know, most of the time, how fortunate we are to have our jobs, and we remember, most of the time, why we love them: for the sheer intellectual stimulation of working with ideas and with works of art; for the diurnal, daunting challenge of teaching and the profound satisfaction of teaching a great class with profoundly satisfied students; for the relative autonomy of our labor conditions, and for a form of labor that is among the least alienated and alienating known to humankind. (Yes, I tell students, it's a 60-hour week, but you get to choose *which* 60.) But we too often forget just what we did to get these jobs, and how conditions have changed since we got them. Semenza, to his credit, retains a visceral sense of all these things, and as a result his book is suffused not with an air of survivor's guilt but with the bracing conviction that both new Ph.D.s and entry-level graduate students need all the help they can get from the people who got that first article acceptance, learned the conference ropes, and wound up on the tenure track.

I can add but one piece of advice to Semenza's guide. It's something about academe that I didn't learn until I had been an assistant professor for a couple of years, whereupon I realized that I had been operating on the principle for almost a decade without knowing it. The principle is this: in this business, as in so many others, you should want other people to trust your judgment. That's basically what "success" comes down to: whether you're writing a seminar paper, refereeing a manuscript for a university press, teaching a class, drafting a committee report, inviting a speaker to campus, or publishing your research, you'll know you've made an impact if your colleagues say, "good call." They can say "good call" in any number of ways—by praising your analysis of *Moll Flanders*, late Wittgenstein, or early Jacksonian democracy; by hiring you on the basis of a fine writing sample and a stellar campus visit; by asking you to serve on a search committee; by asking you to help run the department. But in each case, the structure of the process remains the same: you say X about Y, and person or persons Z evaluate X, which means they evaluate you, which means they evaluate your mechanisms of evaluation. And the more completely those persons Z trust your judgment, the more often you'll be asked to exercise it. Even here, however, as Semenza duly notes, you need to be careful and to pick your spots—lest you wind up on dozens of department, college, and disciplinary committees simply because people know you can be counted on to be a discerning and capable committee member. After all, part of exercising good professional judgment entails knowing those committee assignments you'd be better off without, even as you dedicate yourself to being a good departmental and professional citizen. But with that caveat, the principle holds: the baseline reason for which we praise other people in this business, and for which we try to promote them and their work—whether they're graduate students, junior colleagues, or anybody else—is that we've determined that they have good professional judgment not only about the material that constitutes the basis for their research and teaching but also about the very mechanisms of professional evaluation themselves.

I realize that this is not so much a piece of advice as a piece of meta-advice, but I hope it will help to serve to introduce Greg Semenza's work. And in that spirit, I'll turn things over to him, with these final words of advice to you:

> Trust this guy. He knows what he's talking about, and his judgment is unerring.

MICHAEL BÉRUBÉ
2005

Introduction

Professional development and long-term career planning are no longer optional activities for graduate students in the humanities. Because of a fiercely competitive job market (only one in three Ph.D.s will earn a tenure-track position), college and university officials see few reasons to hire new Ph.D.s who are unable to demonstrate significant publication, research, and teaching records. In light of this fact, we might reasonably ask whether graduate education has changed significantly enough over the past quarter-century to accommodate our graduate students' professional and practical needs. Since an already bad job market has managed to worsen in a relatively short period of time, and since an entire pre-Boomer generation of university professors hangs on the verge of retirement, we should probably confront one of the more troubling and undeniable paradoxes of twenty-first-century graduate education: that MAs and Ph.D.s who must publish, attend conferences, and teach upper-level courses are regularly taught by professors who did none of these things as graduate students and, in some cases, even as assistant professors. While most graduate faculty members surely understand the serious problems facing their students today, there remains a major gap between the lip service often paid to addressing the problems and the implementation of real-world policies and practices designed to alleviate them.

Since I was still a Ph.D. student just 4 years ago, I understand all too well the psychological toll that preprofessional pressures can exact on a typical student in today's academic climate. At a certain point in one's graduate career, simple awareness of what one needs to do to obtain a job can turn to paralysis in the face of having actually to do it. In almost every seminar, you will be encouraged by your professors to publish articles. At every social event, you will overhear stories about the experience of attending conferences and delivering papers. After each semester, you will be forced to ask yourself whether your teaching evaluations are up to par with those of your colleagues. Throughout your graduate career, you will be bombarded by devastating statistics about the job market, many of which will seem custom-made to deepen your own personal anxieties. And despite all of these reminders about *what* you will need to do to succeed, only rarely will someone actually stop and explain to you *how* you might do it.

As mentioned in the preface, this book is different from other graduate school guides in its focus on how to develop an academic career; merely *surviving* graduate school is hardly the goal of most MAs and Ph.D.s. Recognizing the unique problems faced by humanities graduate students, this book seeks to compensate for the inadequate professional training provided by so many graduate programs in the United States and Canada. Unlike other guides, whose authors seem to assume that every reader is a student at Harvard and, consequently, a shoe-in on the job market (which is a bad assumption, anyway), *Graduate Study for the Twenty-First Century* faces head-on the practical obstacles to success for students who will have no obvious advantages on the job market. Because I imagine an audience of recently admitted or already enrolled humanities students, the book does not weigh the pros and cons of attending graduate school, discuss the process of selecting appropriate programs, or deal with how to apply for graduate school. Nor does it spend time outlining nonacademic career options for terminal MAs or Ph.D.s. Whereas several of the existing graduate school handbooks do treat subjects such as dissertation writing and even publishing, their excessive focus on whether and where to go to graduate school also means that they pay insufficient attention to the issues that matter most to the tens of thousands of graduate students who know exactly what they want, having already made up their minds to pursue the MA and then the Ph.D. Rather than teaching you simply how to be a graduate student, then, this book teaches you how to use graduate school as a preparation for what you really seek: a successful academic career.

I want to be honest up front about the fact that this book advocates a sort of "working class" approach to graduate study. Since I pursued my doctorate at a large state university, I was painfully aware as a student that I would need to distinguish myself professionally in order to be competitive in a job market teeming with Ivy Leaguers and Stanford graduates. Ironically, it was this potentially disabling realization that inspired me to keep working. If there's one point I want you to take seriously in this book, it's that whereas the recent emphasis on preprofessionalism can be understood as merely terrifying and damaging, few developments have done more to advance the cause of a more meritocratic system in academe. Now, obviously, we should not ignore the various social factors that continue to condition who goes to graduate school. But, whereas 30 years ago, a state university Ph.D.'s chances of being hired by a major institution would have been limited due to class biases and popular myths about academic pedigree, today's graduate students are more likely to be hired on the basis of their actual qualifications. Potential employers may continue to suspect—erroneously—that a doctorate from Wisconsin is not the same as one from Columbia (is the inference that Wisconsin professors are holding back important information?), but they will have a very difficult time ignoring a Wisconsin student who has published two articles in the best journals in her field. Simply put, professional achievements such as publications and grants can be great equalizers in a rigidly hierarchical and traditionally unfair system. If you regard pressures to develop professionally as merely a burden, you may founder in graduate school; regard them as opportunities for leveling the playing field, and you may go very far.

In case this elite/nonelite scenario seems overly divisive, I want to stress that biases work both ways in academe. The job crisis of the last 25 years has meant that there's no guarantee that top-20 graduates will be hired in top-20 programs; no one in today's academic market, in other words, can simply write off two-thirds of the colleges and universities in this country. Many Ivy League Ph.D.s find themselves being systematically excluded from certain job searches, however, because of unfair assumptions regarding their willingness to profess in nonelite college and university settings. In fact, job placement has become an extremely difficult matter for faculty and administrators at many prestigious universities, where placement rates have in many cases sunk below those reported by institutions usually ranked lower. Whereas the very best students at universities such as Yale and Penn continue to land the most sought after jobs in the country, many Ivy League candidates find themselves in something of a double bind: lacking the professional qualifications necessary to land the most prestigious jobs in the country, they also are shunned by employers at other institutions, who fear that their new assistant professor may bolt for a "better" job at the first chance she gets. Especially for those candidates who wish to teach in small colleges or public institutions—which happen to constitute the vast majority of higher educational venues—such assumptions can be extremely frustrating and very difficult to overcome. Just as students at lower-ranked institutions are sometimes able to research their way into a particular sort of job, these individuals can strengthen their job candidacy by developing teaching and service records reflective of their sincere commitment to the ideals of liberal arts colleges or, at least, less research-oriented universities.

Though I continue to state throughout this book my conviction that preprofessionalism can be regarded as liberating and empowering, I have no intention of downplaying here the dangers inherent in the professional development model of graduate education. Most important, new graduate students should keep in mind that the main purpose of a graduate education is the accumulation of knowledge in an advanced area of study. To the degree that the presentation of conference papers or the publication of articles contributes to your colleagues' or your own understanding of a particular subject, professional activities are wonderfully useful, even crucial, components of the academic life. When they are pursued merely for their own sake—or when the desire of the pursuer to build a certain type of career becomes more important than the desire to learn and grow intellectually—the very integrity of the humanities enterprise is severely compromised. Also, graduate students must be careful to avoid the equivalent of stunting their growth or burning themselves out by trying to do too much, too soon. Although this book suggests that MA students have much to gain by learning early in their careers what is required to become a professor, such students should remember that it will likely take years before the presentation of a paper at a major conference or the publication of an article are realistic goals. The first aim of every graduate student should be to know something extraordinary or at least something ordinary deeply. The second should be to learn how to discuss that subject clearly and persuasively. Only at this point will it be constructive for one to pursue such an ambitious goal as publication. (In chapter 2 of this book, I suggest an ideal timeline for approaching such professional activities). Finally, an overemphasis on professional development can lead to overspecialization, which, in turn, can cause more problems for you on the job market. A very small percentage of universities (about 10 percent) are classified by the Carnegie Foundation as "Research Universities."[1] With a few exceptions, the other 90 percent of colleges and universities tend to privilege teaching and service above research. At many of these colleges and universities, faculties are relatively small; whereas a person writing a dissertation on Shakespeare might only teach Shakespeare at a research university, she would likely be responsible for teaching all English literature through the eighteenth century at a liberal arts college. Students should make it a point to start becoming experts in their respective fields of specialization as early as year one, especially if they plan to pursue a serious research career, but they also should keep in mind the fact that most potential employers are interested in candidates with a broad knowledge of a particular discipline. This book focuses on strategies, therefore, designed to make you as appealing as possible to the widest range of potential hiring institutions.

The tone of this book is direct and, at times, deliberately and systematically provocative. Whereas I am quite willing to meet cultural expectations for rhetorical moderation in my regular academic writing (we all give in, eventually), I've written this book in the voice of a teacher, and I've decided not to edit out what may often seem to you like overly strong opinions. For example, in chapter 4, I offer the following advice to lazy literature students who fail to complete their reading assignments for class: "if you find yourself lacking the energy to read a George Eliot novel on your own, leave graduate school now." While I would defend the sentence here on the grounds that there is, of course, a wider context in which it needs to be understood, it would perhaps be dishonest of me to deny that it's somewhat strongly worded. And yet, as I learned in the classroom years ago, first as a student and later as a teacher, human beings respond to strong ideas and opinions, and they tend to learn extremely little from colorless observations and statements of the obvious. If nothing else, my goal in offering this book is to stimulate serious discussion of issues too often ignored in the course of a graduate education, and so I see no reason to pretend that we will, or even should, agree about all of the ideas it puts forward.

The last thing we need is more deception and dishonesty about the current state of affairs in graduate education. Few, if any, professional commitments are more serious than those made by individuals who embark upon the path to a Ph.D. in the humanities. Rare beings in a society driven by the pursuit of wealth and personal gain, humanities graduate students almost always begin their careers with the most noble of intentions. Since the average time for completing the Ph.D. is 9 years in the humanities, and since many graduate students accumulate significant debt during that time (debt that will not be easily paid off on a professor's salary), and many others won't be hired on the tenure track, it is incumbent upon all in higher education to review current practices and policies.[2] The sort of institutional dishonesty about which I'm speaking only rarely takes the form of outright lies; more often, it amounts to a refusal on the part of administrators and faculty to address the practical needs of their students.

Specifically, too many faculty members continue to treat their students as mere "apprentices," despite the fact that graduate students in most modern universities design and teach their own classes, serve on departmental and university committees, and conference and publish regularly. The error is somewhat understandable, but not entirely excusable. Pressure to maintain the traditional "apprenticeship" model of graduate education is imposed mainly from above, since high-level administrators and university attorneys, determined to prevent graduate student unionization and thereby maintain an increasingly massive and inexpensive labor force, require and advise that teaching and research assistants be classified as apprentices, not professionals. To refer to an individual who is thrown into a classroom with little advanced training on the first day of her graduate career as an "apprentice," however, is to redefine rather completely the meaning of that term.[3] To say that students who must publish prior to graduation are "apprentices," for example, is to imply that we will actually educate them about the publication process as it pertains to academic journals, and university and trade presses. The simple fact that tends to get lost in the confusion of university politics and corporate economics, however, is that graduate programs not only admit annually far more students than the market can accommodate—and for all the wrong reasons—but also that they do painfully little to prepare these students for the realities of academe in the twenty-first century. The costs of these lies are reflected partly in the numbers. A *Chronicle of Higher Education* cover story from January 16, 2004 reveals that attrition rates in U.S. Ph.D. programs are at an all-time high, between 40 percent and 50 percent (higher for women and minorities).[4] Above all else, the statistic highlights waste of time and resources by universities and, more important, of money, time, and energy by graduate students. Such numbers speak to the general feelings of alienation and aimlessness experienced by so many graduate students. And they speak to the general failure of universities—faculty members included—to take adequate responsibility for their students/employees. As Michael Bérubé and Cary Nelson have argued, "Faculty members who devote no energy to graduate training have a relation to graduate employment that is almost wholly parasitic: their own salaries and privileges are sustained by exploiting teaching assistants."[5]

So let's be honest for a moment and consider the vicious cycle that's producing the current crisis in graduate education. Universities admit annually more graduate students than the market can accommodate in an effort to staff their undergraduate classes. Whereas in the past, most of these classes were taught by tenured or tenure-track faculty members, university officials eventually caught on that graduate students and adjuncts could do the same type of work for far less money and few, if any, additional benefits. Further, because neither graduate students nor adjuncts have tenure (i.e., academic freedom and job security), they represent a workforce that can be easily managed and manipulated by their employers. Since the late 1970s, the percentage of full-time tenure-track faculty members has steadily decreased as graduate students, adjuncts, and part-time faculty members have taken over their duties. In fact the U.S. Department of Education reports that since 1981—a period during which the population of college students has grown exponentially—

the percentage of full-time faculty members has decreased from 65 percent, which is bad enough, to only 51 percent.[6] So get this: in the past 25 years or so, universities have systematically reduced tenure-track lines by replacing tenurable professors with "apprentices" who seek nothing other than to be hired on the tenure track! An additional irony is that administrators and state legislators have few incentives for addressing such problems as Ph.D. attrition rates since attrition is precisely what keeps the job market from becoming more flooded than it already is. Perhaps most troubling, though, is the very real threat posed to academic freedom as tenure-track jobs continue to disappear in both our public and private colleges and universities.

The only realistic long-term solution to this national, systemic problem may be graduate student (and adjunct) unionization. More than 30 graduate student unions are currently recognized as collective bargaining agents by their universities and state or federal legislatures; at least 20 others have recently affiliated with unions and are in the process of seeking legal recognition as collective bargaining agents.[7] While many conservative commentators and university administrators continue to argue that graduate student unionization will lead directly to the downfall of higher education in the United States, basic common sense and numerous historical precedents have suggested precisely the opposite: the superior wages and benefits earned by members of graduate student unions promise at least two positive side effects: first, by raising the costs of graduate student labor, they force universities to think twice about admitting too many applicants, who will then flood the job market a few years later; second, by limiting the financial benefits of hiring graduate students rather than assistant professors, they slow down and may eventually help to prevent the current corporate assault on tenure. Unless one is able to claim with a straight face that unionized students in such prestigious graduate schools as Berkeley, UCLA, Michigan, NYU, Rutgers, and Wisconsin (with the first union, organized in 1966), seem to be struggling as a result of having unionized, arguments against the move to protect the rights of a badly exploited labor force seem totally unpersuasive and unethical. Even in cases where union movements have eventually failed, activist graduate student bodies have tended to benefit from the concessions offered by their universities in their attempts to block unionization. "If we can't beat them into submission, we can at least pretend to treat them fairly," would appear to be the line taken by many administrations. The unionization issue is undoubtedly complex but, as a graduate student in today's market, you should at the very least make it a point to become educated about the major issues pertaining to the unionization movement. Faculty members, regardless of how they feel about unionization, should stress to their graduate students that they have a right to organize and that reprisals from either the department or the university are illegal (and not in the best interests of anyone). The eventual fate of the graduate student union movement will have serious implications down the road for everyone involved in higher education.

One popular, alternative method for addressing the graduate student job crisis amounts basically to a Band-Aid where a tourniquet is needed. I'm talking about attempts by departments to offer job training for Ph.D.s who decide, almost always out of desperation, to seek employment in nonacademic professions. While workshops on nonacademic employment opportunities seem like a nice idea—and shouldn't be *discouraged*—we should be honest about the fact that they serve the sole purpose of cleaning up a mess that should have been prevented in the first place. I am quite willing to wager that no Ph.D. student enters a program in the humanities to become an editor, a freelance writer, or a lab technician.[8]

As always, more innovative educational initiatives may be the only practical solution for today's graduate students—but not in the sense that they will make the larger problems we've been describing go away any time soon. Speaking realistically, unless the increasing corporatization of the academy can be halted, and unless graduate students and adjuncts can win the right to bargain collectively in both public and private university settings, the problems are unlikely *ever* to go away. By educating yourself about how the current system works, however, and seeking to reform (mainly non-curricular) departmental practices so that your professional needs are met more effectively, you can at least maximize your chances of success in the current market. As a sincere believer in the idea that cream, if given the chance, still will rise to the top (even in this awful market), I offer in this book the information I believe you will most need to know in order to excel as a future professor in the humanities. Here's that much-needed apprenticeship, in other words, that you may find lacking in your department. And though the advisory system I lay out here may seem at times cold or mechanical, it comes from a deeply personal place since it's exactly the sort of plan that saved me.

Before college, I was a god-awful student—bored, lazy, and a bit hostile. My grades stayed in the upper "C" range. I continued to flunk math. I had a lot of fun and somehow kept getting by, but I had no vision whatsoever of ever living outside of my small, working-class New Jersey town. I was pretty good in English only because I read voraciously as a kid. I wrote a lot too—poetry, short stories, plays. All mediocre pieces, of course, but they managed to catch the attention of two English teachers in my small Catholic high school who, somewhat inexplicably, believed in my ability to "turn things around." They read my poetry and talked to me about it. They explained to me that I could actually study literature in college and suggested things I might do with an English major. They saw something in me, or at least they pretended to. So without asking many questions, I took their advice my senior year and applied to a few strong creative writing programs. The most attractive one was the University of North Carolina at Greensboro. I requested an application for the college in Greensboro, mailed it in, and was quite surprised a few months later to learn I had been accepted. Considering how careless and ignorant I had been about the whole process, it seemed too good to be true.

That spring my freshly divorced parents and I drove down to Greensboro for a freshmen orientation weekend and my first chance to see the campus. When we pulled in the gates marked UNCG, I felt relieved, happy even. The excitement ended when we were informed by a smirking security guard that the "West Residence Hall" to which I had been assigned was not, in fact, located on the UNCG campus but could be found instead on the campus of Greensboro College, which was just down the road to the right. I had applied to the wrong school.

Greensboro College is a very small Methodist college of about a thousand students—nothing like UNCG. One cannot get lost on its campus. Its library is not large. Its total acreage is comparable to that of my high school. The moment I was alone in a bathroom on the right campus I just lost it. I was angry about the lack of guidance I had received, but it was impossible to deny my own stupidity and carelessness. The indifference had finally caught up with me, it seemed. My parents tried their best to cheer me up and negotiate their own embarrassment, and I put on a happy face and got through the weekend. But once I got back home, too ashamed to tell anyone what had happened, I sunk into a rather severe depression, not yet having even the slightest clue of Greensboro College's importance in my life.

In September, I went crazy. I attended about half of my classes, barely did any work at all, and got as high as possible as often as possible. I also succeeded in completing my first philanthropic group project, which involved several of my friends and I running a small fake-ID business out of a third floor dorm room. A few weeks later, three of those friends and I got busted for carrying paraphernalia and a small amount of weed. I figured it was all over at that point since getting kicked out of college for drug possession doesn't look particularly good on a resume.

I'll never know why, but the dean gave us all a second chance. Granted, we were scared and sincerely repentant, but she still didn't have to be so humane. Instead we were put on probation, required to attend regular drug counseling sessions, and told we had the rest of the semester to shape up. I was terrified. I was pretty sure that if my father found out, he would kill me. I knew that my girlfriend would dump me. And so with a reason suddenly to do so—with no choice, really—I studied. The small college environment was nurturing, and my professors were dedicated to teaching and supportive of their students' well being. I finished the semester with a 3.8 GPA and enough newfound confidence to pull straight A's in the spring. I transferred to Rutgers the following fall and fell in love with Shakespeare. Ten years later, I was hired as an English professor at UConn.

I suspect that your path into academe was different. My reason for sharing this personal history, in fact, is to say that it absolutely doesn't matter how you get here. If you're reading this book because you wish to become a professor, it just matters that you eventually do. I feel compelled to make this point because I think the worst way to read this book is to regard the advice it offers as prescriptive of a "right way" of doing things. I've been pleased, of course, by this book's positive reception since 2005, but I've come to feel that where it tends to be most specific it invites the charge of seeming dogmatic and rigid. This was not my intention. As I remarked back then, "I see no reason to pretend that we will, or even should, agree about all of the ideas [this book] puts forward." I wish to reemphasize the point here in order to clarify my conviction that more formal rules of behavior, and more academic socializing practices, are the last things we need. I hope that you will find the information in this book extremely useful, but I also hope you will adapt it to your own personality. Know what you're facing in the years ahead, learn what you need to know about how to deal with it all, and then apply your newly acquired knowledge to your own work style and personal habits.

I recommend that you seek out, early in your career, absolutely every bit of professional advice you can, from every experienced academic willing to offer it. Working mainly from my own experiences and with a profound sense of gratitude toward those advisors who helped me along, I originally wrote *Graduate Study for the Twenty-First Century* because I was dissatisfied with the systematic blind spots of graduate advising. In many ways, the book suggests that you must advise yourself and indicates the crisis in advising that many students experience. While I had enjoyed challenging seminars and benefitted from extraordinarily generous advisors, I also felt frustrated much of the time by having had to complete a so-called "apprenticeship" with so little practical training. Information about how to read texts carefully was plentiful; information about how to engage an academic audience, how to prepare a manuscript for publication, or how to field conference questions well was sorely lacking. I walked around much of the time feeling ignorant. I thought my professors were intimidating because they seemed so together all the time, and I assumed erroneously that they had always seemed so—that they had just *known*. But no one just knows. Even the children of academics, even students with the "perfect" pedigree, have to learn the ropes at some point.

The only difference between you and your professors, generally speaking, is their relative experience—the fact that they've somehow already gotten where you're going. The more quickly you arrive at the realization that your professors are human beings, who bumbled and now are bumbling through their busy lives with most of the same anxieties and uncertainties as you, the more quickly you will feel empowered to perform the work, ask the questions, and establish the sort of self-discipline that will allow you to thrive as an academic. Continue to view academic success as the result of superior intelligence or innate know-how and you will probably struggle as you move up the ranks; learn to see it as the result of hard work and the accumulation of practical knowledge, and you will find it easier to excel. This is not to say that you will, or ever should, feel like you've got it all figured out.

I remember quite vividly—since we always remember our *lowest* moments—asking one of my graduate professors how academics are to measure the value of what they do for a living, a question I still ask myself on an almost daily basis. It was a question that emerged, appropriately I should point out, from our discussion of Dr. Faustus' dissatisfaction, even boredom, with the fields of logic, medicine, law, and theology—his frustration, that is, with the limitations of traditional academic learning. Eventually, Faustus decides that the academic life is too unsatisfying, and he sells his soul to the devil in exchange for *real* answers to all of his previously unanswerable inquiries. When I asked the question, my professor cast upon me a look of unspeakable fierceness and claimed he couldn't "begin to understand" what I meant since he had no such doubts about what he did, said that he imagined most successful academics had no such doubts about what they did, and then, changing the subject, asked the class what we were to make of Faustus' turn away from academics in favor of black magic. Like Faustus, I was a graduate student seeking metaphysics, and all I could get was bitter "analytics." And so Faustus's question became my question: can this profession afford no greater miracle than *this*?

I think the anecdote demonstrates what we might call very generally a failure in graduate advising, a skill few young academics are ever asked to think about, either while completing their degree or working towards tenure. You might argue that my professor was acting in the capacity of a teacher rather than an advisor since the exchange occurred in a classroom. But one of the great complexities of advising is that it isn't defined by place in quite the same manner as the other major academic activities. Generally speaking, research happens in the library or the lab or the office. Teaching usually happens in the classroom, and it spills over into office hours. Service happens in committee rooms and, I'm guessing, also in hell. But advising happens everywhere and it happens without much planning, and it cannot be easily distinguished either from teaching, service, or research. This is one reason why even though it probably should be regarded as the fourth of the major academic activities—categorized right alongside the holy trinity of teaching, service, and research—advising is not discussed much at all, especially in forums where faculty and graduate students are together.

But I have another motive for sharing this particular anecdote. Even though my dedicated and typically helpful professor was quite willing in many cases to digress from a strict analysis of assigned works—for example, he spent hours over the course of the semester talking to the class about his various works in progress, about his literary tourist experiences, etc.—in this particular case, he clearly was intent on sticking to the text. Did his response to my question, I wondered, suggest anxiety about what it would mean to acknowledge possible flaws in the academic system, would sharing his own doubts reveal a personal vulnerability that would undo the ethos he had worked so carefully to construct? Would indulging a brief discussion of professional-development-related issues represent a dumbing-down of the conversation—a turn away from the serious intellectual *content* of the class to the baser, practical side of the profession that no *serious* scholar would ever really care about?

These are loaded questions, obviously, intended to get at what I see as a tendency in higher education to regard graduate study and graduate professionalization as mutually exclusive or, worse, as conflicting categories. Simply put, scholars and professionals aren't always allowed to mix. For example, Jeffrey Jerome Cohen, the eminent Medievalist and Chair of English at George Washington University, recently published on his blog his orientation chat with entering GW grad students. So you can accurately understand his position, I quote him at length:

> Graduate school in the humanities is one of the few arenas available in the United States where you can be an unabashed intellectual, where you

> don't have to justify the research you are pursuing solely in terms of immediate use value or according to some cold economic calculus. I urged the gathered students to think of themselves as already a part of a community, to attend every lecture and symposium that they could find the time to attend, to challenge their professors and each other in their seminars, to never be ashamed of the enthusiasm that motivates them, no matter how abstract, abstruse, or even grandiose their preoccupations. Mostly, I urged them to take what they could from the experience of being in graduate school without obsessing to an extreme over what comes next (I believe too many graduate programs overemphasize professionalization, so that students become neurotic about conference papers and publication—as if there were some magic checklist that when completed yields a first job). I know it is easy to declare, having been fairly successful at this career, but even if I had never landed in that first tenure track job, I wouldn't have regretted my time as a graduate student.[9]

This is a thoughtful, nuanced entry, but with a few debatable points. On the one hand, Cohen reasonably stresses that he is opposed to students "obsessing" over what comes next, and he's right, of course, that there's no "magic checklist" that when completed will yield a first job. There is, though, a *real* checklist that when completed can increase exponentially a student's chances of getting a first job. Let's consider for a moment the type of student Cohen is referring to: an English graduate student who wishes to become a professor: now, a candidate who goes on the market with two articles, four or five conference presentations, and high scores on teacher evaluations will be far more likely to get a job than one who has not established such professional credentials. As another successful academic blogger suggested in response to Cohen's advice, telling grad students not to worry about professionalization is "a bit like telling a farmer not to worry about his newly planted crops. Sure, they've *just* been planted, but they don't get to be fully grown, ripe-for-the-picking, healthy crops on their own."[10] Further, what would it mean for a graduate student to "obsess to an extreme" about professionalization anyway? Presumably Cohen is envisioning a focus on professional development so intense as to cause either paralysis in the candidate or intellectual shallowness, but as I argue above, most graduate students worry more about not knowing *how* to professionalize than about the simple fact that they *have* to professionalize. No amount of frustration about reality is going to change reality; graduate students are smart people; they know what they're facing, and they want the tools they need to navigate the future. Perhaps Cohen is sincere when he says that graduate students should be happy just being in graduate school and not worrying about getting a job, but I can say to you unequivocally that after six years in graduate school and the $25,000 of debt onto that I was forced to accumulate as an undergraduate student, I certainly would have regretted my time as a graduate student had I never landed a job.

What links Cohen's and my Marlowe Professor's comments is a false dualism often drawn between impassioned intellectualism and professionalism. Here we have a pastoral image of graduate education: students frolic on hillsides to the soothing sounds of oaten pipes, an infinite supply of books on hand, inspiration spontaneously charging through their veins until it overflows the fields in the form of powerful emotions. It is idealistic, even nostalgic, and it effaces labor concerns and market realities with surprisingly little embarrassment. It also is undeniably attractive since we all long on some level for the sort of intellectual environment that it evokes. In any case, a rift continues to divide those who believe that advising should be about how we seek to position graduate students within an idealized intellectual setting and those who believe it should be about how we seek to position them within a more realistic one. I am inclined to call those persons who believe it should be about *both* "good advisors"—plain and simple, and I believe that the future of advising practices in graduate education must focus on closing this rift.

In order to accomplish this, we need perhaps only focus on our own history. Although the *OED's* primary definition of an "advisor" is considerably vague, "One who advises or counsels," the second has all kinds of interesting implications: "Chiefly *US*. At some universities, a senior member assigned individually to advise students on personal, academic, and other matters. Cf. *moral tutor*."[11] The first recorded usage goes back to 1887 when a writer in *Lippincott's Monthly Magazine* wrote the following:

> One great power of appeal . . . playing between teachers and students [at Harvard] is exercised through the "advisers." Each matriculate is expected to designate one of his professors whom he will consider his adviser while at the university. The professor is to be consulted by the student as a personal friend and guide.[12]

The striking features of the entry and example include the description of the advisor as a "senior member" of the faculty; in most universities today, junior faculty members are permitted to serve as advisors. Even as early as 1914, a first-year Princeton student would have been appointed an "adviser," described as a "younger member . . . of the faculty to whom [the student] is encouraged to go with all or any of his perplexities."[13] "Perplexities" is another wonderfully vague term, but one assumes that it refers to something like the range of problems described in the definition itself as "personal, academic, or other" matters. Finally, there are the intriguing suggestions that an academic advisor is similar to either a "moral tutor" or a "personal friend." While I cringe at both "moral" and "friend," I'm struck by the degree to which advising here refers to a relationship between teacher and student that far exceeds the boundaries of the curricular.

Little research exists on the graduate advisor/advisee relationship, and most of what has been published comes out of the social sciences. All of the published data, however, backs up the idea that good advising is holistic, and I'd like very briefly to report some of its basic conclusions. First, surveyed students regularly claim that the health of their advising relationships ranks very high as a determinant of one's success in academe: as Schlosser and Gelso write, "negative impacts on students have been found from negative mentoring or from a lack of faculty mentoring. Results from . . . empirical studies have pointed to a need in the literature to develop a greater understanding of the advising relationship."[14] Across all of the studies underlined by this type of claim, we find much consistency: in "A Qualitative Examination of Graduate Advising Relationships: [from] the Advisee Perspective," published in 2003 by Schlosser, Hill, Knox, and Moskovitz, the data reveal those specific factors influencing whether advisees view their advisory relationship as "satisfying" or "unsatisfying."[15] Unsatisfied students report that "career guidance was typically not a part of advising relationships."[16] Conversely, "satisfied students indicated that their advisors encouraged them to participate in professional conferences and/or introduced them to important people, [etc.]."[17] Further, whereas "Satisfied students typically reported feeling very comfortable disclosing aspects of their professional lives to their advisors," "Unsatisfied students . . . typically reported feeling cautious talking about their professional lives."[18] Finally, whereas "satisfied students indicated caution about sharing personal information with their advisors," "All . . . unsatisfied students indicated being cautious sharing personal information."[19] I quote the overall findings of the 2003 study:

> In sum, the positive advising relationship could be described as one in which the members have a good rapport, process conflict openly, and work together to facilitate the advisee's progress through the graduate program and development as an emerging professional. . . . Conversely, students who are unsatisfied with their advising relationships . . . are unlikely to refer to their advisors as mentors, because the term mentor connotes a positive valence.[20]

In an earlier study by de Valero (2001) into how advising factors influence time to completion rates—infamously longer in the humanities and social sciences—"the style of advising [is shown to be] another important factor for graduate student success."[21] De Valero points out that studies going back as far as 1992 reveal that "the most common advisors were those who allowed students to work at their own pace, without establishing any work schedule and timetable. . . . [T]his style of advising is dangerous, because some kind of schedule must be imposed to periodically monitor the research work. . . . High stress levels and pressure to complete the dissertation are major factors in graduate student attrition, particularly when students do not have a close relationship with the chairs of their committees, when they work alone, and when less feedback is provided."[22] According to de Valero, in those departments able to boast of the lowest attrition rates and shortest times to degree, "the kind of words [students] used to describe [the advising relationship] differed from those used in other [departments]. The most common words used to describe positive student-advisor relationship[s] were: 'excellent,' 'nurturing,' 'mentoring,' 'caring,' and 'exceptional.'"[23] In yet another 2001 study measuring the "Working Alliance" between advisors and advisees, "Higher scores . . . indicate an alliance where there is a strong interpersonal connection, and the advisee feels respected, encouraged, and supported . . . [and] teaches the advisee how to function within the profession."[24]

I want to stress that concepts such as nurture, interpersonal warmth, and even terms such as "loving" and "caring," are never equated with, nor used to promote, what one might equate with "easiness" or "touchy-feeliness." This is due largely to the fact that students' "respect for the advisor" and highly organized expectations management by the advisor tend to be crucial factors in positive relationships. When students say that they want a friendly relationship with their advisors, they don't mean they want to go out and get drunk with them. One student represents her colleagues when she says, "I feel very comfortable with [my advisor] and I don't mean warm and fuzzy . . .; I feel comfortable expressing disagreements to her, and I know when disagreements come up, we are able to bring it to the table and talk about it."[25] The pervasive idea through much of the literature, then, seems to be that regardless of an advisor's particular personality traits,[26] her ability and willingness to connect with the advisee intellectually, personally, and professionally is what makes her a successful advisor. By demystifying the academic process, faculty advisors give their students a greater sense of control and direction.

Graduate students are often told that they should be grateful for the opportunities they've been given, and I wouldn't disagree with the basic point; but I also believe that it's occasionally necessary to remind faculty and administration why we should be thankful for those opportunities too. As Bèrubè and Nelson have pointed out (see above, p. 6), faculty members are able to enjoy the time they need to pursue their research agendas while benefiting from enviable teaching loads made possible and sustained only through the labor of graduate students. The least that universities can do, I think, is everything they can to ensure that their students obtain the best education possible while preparing them for the realities of the academic marketplace, and the pressures, anxieties, and questions they are likely to have about living a life in the highly idiosyncratic world of academe. As a faculty member, I continue to ask all sort of questions about the value and the usefulness of the things I do on a daily basis, but the importance of advising graduate students—and the ethical obligation I have to do it thoroughly—is something I never doubt. My hope is that you will find the professional advice offered in the following pages illuminating and helpful. Learn it, adapt it to your own disciplinary and personal practices, and use it freely to get wherever you wish to be going.

Graduate Study for the Twenty-First Century is organized into 12 chapters that cover the graduate experience from the first seminar to the first job. While you may be tempted to skip directly to chapters that you assume will be most relevant, I would encourage you to read the entire book in the order that it is presented. Because the book seeks to explain the vital connections between each stage of the process, highlighting especially how one particular phase or activity can be used as preparation for the next, later chapters are less useful in isolation. For example, although Chapter 10 focuses on "Publishing," it builds directly on ideas presented in Chapter 1 (on the publication industry and pressure to publish), Chapter 2 (on when to publish), Chapter 3 (on time management strategies that make publication easier), and Chapter 5 (on the research process). By the time you finish this book, you should understand where all of the pieces of the puzzle belong; then it will be your job to put them together.

I want to close by commending your decision to pursue an academic career. Even all these years later, I still believe there are few jobs more important or fulfilling than a university professorship in the humanities. The issues touched on here have, in fact, made even more apparent the crucial role in our culture of dedicated intellectuals and educators. As a modern graduate student wrestling with modern problems, you'll need to fight harder than most of your academic predecessors ever had to do in order to keep in mind the heroic nature of the enterprise upon which you've embarked. If you take no other advice away from this book, I hope you'll at least remember to keep faith in the transformational power of humane knowledge.

CHAPTER 1

THE CULTURE OF A GRADUATE PROGRAM

Few undergraduates know or care all that much about how their major departments operate and, in truth, their ignorance probably has no negative consequences. Only rarely are they ever invited to participate in the administrative or curricular management of a department. To succeed in graduate school, however, students must learn quickly about how academic departments—and the individuals who run them—are organized and governed. In the worst cases, ignorance about such factors can lead graduate students to act in ways extremely damaging to their reputations and careers. Based on the premise that both successful graduate study and professional development begin with an understanding of academic culture per se, this chapter provides nuts-and-bolts information on a variety of general subjects, including:

- The daily life of a typical humanities professor
- The tenure and promotion system
- The hierarchical structure of a typical department
- The major characters in an academic department
- The politics of academic life
- The intensity of graduate study

By describing in relatively concrete terms the undeniably complex habitat of the humanities scholar, the chapter aims to make you confident in your ability to participate fully and "safely" in the life of your department.

THE DEPARTMENTAL HIERARCHY

Keeping straight all of the people in a university department often proves a job in itself at the start of one's graduate career. Though faculty members and departmental administrators must deal regularly with the "higher-ups"—the presidents, provosts, and deans of colleges—graduate students need only rarely involve themselves in extra-departmental affairs (this is a fact, not a recommendation or endorsement) and are unlikely to have much contact with such individuals. Even though some people love to talk about the university as an ideal, democratic space, removed and free from the corrupt practices and structures of the business world, the fact is that academe has in recent years become nearly as corporatized and hierarchical as a typical Fortune 500 company. And like individuals working in the business world, academics need to understand the ways in which power is distributed, exercised, and balanced if they are to enjoy successful careers. Here's how things are typically organized.

Administrators

Departments are directed either by a "Head" or a "Chairperson." Technically, the difference is that whereas heads usually are appointed by the dean of the college, chairs are typically elected by, or at least supposed to be representative of, the faculty, though I should mention that lots of departments use "head" for either form of government. The implications of the appointment process can be quite serious, as you can imagine, since that process potentially defines the difference between autocracy and democracy in a given department. In most cases, a wise department head will try to represent the majority of faculty even though the dean happens to be his official boss. In many departments, "executive committees" are set up to advise the head or chair and, depending on how their role is defined, balance the power of the department head/chair. Some heads/chairs involve themselves directly in the governance of the graduate program, and some choose to grant near-autonomy to the director of graduate studies. Over the course of your graduate career, you may actually have very little contact with your head or chairperson, but you should at least make sure that he knows who you are.

The "Associate" head or chairperson is both an advisor and a supervisor of certain important administrative tasks such as the scheduling of undergraduate classes, the distribution of graduate teaching assignments, and the hiring of adjuncts. The associate head/chair may also serve *ex officio* on any number of departmental and college-level committees, including courses and curriculum and the department executive committee.

The "Director of Graduate Studies" is very likely to be the highest-ranking administrator with whom you will work on a regular basis. The director is responsible for establishing graduate course schedules, issuing exams, training job market candidates, managing the graduate admissions process, and overseeing each student's progress through the MA and Ph.D. programs, among other things. There is considerable disagreement in academic circles about how the role of the director should be defined: as taskmaster, confidante, or something in between these two extremes. On the one hand, the graduate director's job is to be an advocate for you; on the other hand, he is an officer of the institution. You probably will be able to intuit upon meeting your director what sort of relationship yours will be. Always remember, though, that graduate directors are not appointed simply to field complaints, though this is a part of their job; you should go to your director to seek advice about everything from how to succeed in seminars to how to survive on the job market.

Faculty

Generally speaking, there are three ranks of professors in most universities. Though a full professor obviously is more highly ranked than an assistant professor, I begin here with the latter in order to emphasize the promotional movement upwards. But first, a quick word about terminology: after the dissertation defense, a Ph.D.'s friends typically begin calling him "Doctor," which is officially accurate only after the degree is conferred. The word "professor" is used to describe persons contracted officially as full, associate, assistant, or visiting professors by colleges or universities.

Assistant Professor: "Assistant Professor" is the utterly inappropriate term ("beginning professor" would be more accurate) used to describe professors who have yet to be tenured or promoted to the associate rank. In most cases, assistant professors are recently defended Ph.D.s who don't really ever "assist" with anything. Having successfully conquered the job market, these individuals sign onto a six-year long trial—the so-called probationary period—during which they are expected to teach a normal course load (though sometimes it is reduced), conduct research, and serve on departmental and university committees.[1] At the end of this probationary period, the candidate will submit a complete file, which will be reviewed at various levels within his own university and by approximately four to eight peer reviewers outside of it (in some elite universities, files are evaluated by as many as twenty outside reviewers—a truly absurd practice). If the candidate has performed his responsibilities satisfactorily, he will be granted tenure.

Tenure decisions are made by a number of departmental and university committees set up to check and balance one another.

In most cases, the initial recommendation for tenure is offered by the promotion and tenure committee, which consists of about five or six tenured members of the candidate's own department. These committee members are responsible for reading through and discussing each candidate's file—including the external reviewers' evaluations—and determining whether or not the individual should be granted tenure. Once the committee offers its recommendation, it must also be approved by the department head/chair and sometimes must be voted on by the faculty at large. After the department approves of a particular candidate's case, it's time for the higher administrators to weigh in on the matter. First the dean (and Dean's Committee) of the college, then the provost or chancellor, and finally, the Board of Trustees all must evaluate the file, which can be rejected at any stage in the process. The most important stamp of approval comes from the dean's committee since, in most cases, these are the people actually responsible for firing people. Depending on whether this committee is representative of a humanities college, a liberal arts college, or a liberal arts and sciences college, its members will come from more or less different academic backgrounds and disciplines. College of liberal arts and sciences committees can be problematic since they subject candidates' research to evaluation by science professors, who often have an inadequate understanding of how research in the humanities should be judged (science candidates are subjected to the same unfair evaluation by humanities professors, of course). Only when the file is approved at all the highest levels will the candidate be granted tenure.

In public university settings, starting assistant professors in the humanities earn a salary of approximately $57,000 in public universities and $62,000 in private ones.[2] Considering the superior educational background of a Ph.D., most outside observers would be shocked, of course, to learn of the disproportionately modest salaries paid to professors; while I certainly wouldn't argue with them, I would stress the importance of recognizing that professorial jobs do include other financial perks. Academic jobs almost invariably bring with them excellent health benefits and competitive retirement plans. In many institutions, additional benefits include money for travel to conferences and archives, opportunities for teaching and research grants, and excellent child care facilities and benefits. Furthermore, a professor's earning potential can be quite good at a competitive university; especially where faculties are unionized, merit pay opportunities can result in significant and regular salary increases. Just as important, it goes without saying that college towns and academe in general tend not to attract budding entrepreneurs and yuppies; a professor would likely feel embarrassed, rather than proud, driving a Lexus to his 9 AM class on the Roman Empire (beat up Volvo's, though, seem to be okay). In short, assistant professors can live quite comfortably in most locations on a typical academic salary.

Associate Professor: Upon being promoted and/or tenured, our professor will join the "associate" ranks. Associate professors form the most diverse constituency of faculty members in most departments, ranging in profile from recently tenured thirty-somethings to professors on the verge of retirement. For their contributions, associate professors earn approximately $67,000 in public universities and $74,000 in private ones. After producing a major new professional credential such as a prominent second book, an associate professor can submit his credentials for promotion consideration.

Professor: The rank of "Professor" is used to designate those individuals who have earned the highest level of distinction in their respective fields. While some exceptional academics earn the title in their mid- to late thirties, most full professors are seasoned veterans of university life. While some continue to involve themselves in every aspect of university and department life, others scale back their service activities in order to focus on research or teaching. Those who stay involved often serve on the college- and university-level committees, perform most of their institution's work on promotion and tenure committees, and also tend to do the bulk of external reviewing and article refereeing in their respective fields. For their pains, professors earn on average $90,000 in public universities and $115,000 in private. Partly for this reason, many professors will work well past the average national age for retirement. While full professorship may seem like a very abstract fantasy to you right now, the title should represent the ultimate achievement for any up-and-coming academic.[3]

Staff

Depending on the size of the department, staffs can be very small (one or two individuals) or include many people. I discuss here only three of the more common types of staff member in most humanities departments.

Contingent Faculty: The modern corporate university is built largely on the backs of its adjuncts and nontenure-track (part-time and full-time) faculty.[4] Whereas adjuncts, many of whom have earned Ph.D.s, are hired to teach on a per-course or hourly basis, usually without any health benefits or retirement options, nontenure-track faculty members usually are salaried workers. Many large research universities employ hundreds of adjuncts, and they pay them extremely poorly for their hard work. Approximately 70 percent of adjuncts in the United States make less than $3,000 per class.[5] This means that just in order to make $30,000 annually, let alone pay for health benefits and childcare, the average adjunct must teach ten courses a year. As bad, because adjuncts are hired on a per semester basis, they can be fired or have their employment discontinued at the drop of a hat. Despite the appalling conditions under which most adjuncts work, the response of the professoriate (and various professional organizations) has been defined mainly by silence and indifference. Many academics who lack a sufficient amount of empathy also fail to consider pragmatically how the abuse of adjuncts is contributing to the gradual winnowing away of tenure at many colleges and universities. This crisis of academic freedom has been exacerbated recently by universities' increasing reliance on part- and full-time nontenure-track faculty. Since 1998 alone, the number of nontenure-track, full-time faculty positions has grown by about 40 percent; such rapid growth helps to explain why more than 60 percent of all faculty appointments in America today are nontenure track—a 40 percent increase since 1988.[6] In the best cases, such workers are hired on six-year contracts, but they often teach as many as four or five courses per term for two-thirds the pay of tenure-track faculty members. For graduate students specifically, the exploitation of adjuncts and nontenure-track faculty members (and, of course, graduate TAs) also means fewer tenure-track jobs will be available once they've finished their Ph.D.s. Cary Nelson and Stephen Watt have rightly referred to the abuse of adjuncts as "the single worst problem higher education faces" because it is "linked to every other crisis in the industry."[7] As a graduate student, you can begin to address the crisis by urging your state legislators and appropriate professional organizations (MLA, CAA, AHA, and APA, among others) to address the problem and by supporting contingent faculty activism on your own campus.

Graduate Teaching Assistants: The term "teaching assistant" is often a misnomer since graduate instructors in many humanities departments regularly plan and teach their own courses. Despite the obviousness of this point, most universities insist on the term because it connotes a supportive role rather than an independent one, which is crucial to the legal classification of graduate students as "apprentices" or "students" instead of "employees." As I discussed in the introduction (pp. 5–7), such classifications serve as a hindrance to the formation of graduate student unions, which force universities to bargain collectively with their graduate and research "assistants," who in many places teach a surprisingly large percentage of undergraduate classes. Graduate students who have organized successfully usually banish the term "assistant" in their first union contracts, opting instead for more accurate terms such as "graduate employee" or "graduate instructor." Like the exploitation of adjuncts, an overreliance on graduate student labor is not only unethical but also has serious practical consequences for the professoriate and the tenure system. Why would any university hire a new tenure-track professor for the same price as three or four graduate instructors? While the answers may be obvious to any educator or supporter of tenure and the academic freedoms that it protects, many politicians, administrators, and trustees clearly are thinking more in terms of dollars and cents than ethical or educational principles. Again, I would urge you to educate yourself about the national controversy regarding graduate student unionization.

Administrative Assistants and Secretaries: Imagine having to deal all day, every day with the eccentricities of intellectuals and the complaints of students in return for a small salary, little job security, and few benefits. In my mind, the various assistants and secretaries in academic departments are the unsung heroes and heroines of colleges and universities. They manage to coordinate an extraordinary number of people and somehow facilitate the smooth operation of massive bureaucracies. Most professors and students are wholly unaware of how absolutely dependent on such staff members their departments happen to be.

Simply put, few relationships will be more important to you than the ones you develop with the department secretaries in the main and graduate offices. If you are rude and dismissive to them, you may find it extraordinarily difficult to get what you need. You will also go a long way toward damaging your reputation since these staff members know and regularly talk to just about every single person in the department. If you are respectful and considerate, on the other hand, your professional life is likely to be considerably more pleasant and efficient than otherwise.

The Life of the Scholar

In a well-reported and now infamous case from 1997, a University of Washington professor set out to mow her lawn in the middle of the morning; an angry witness to this despicable outrage—determined that his tax dollars should be better spent—reported the incident to horrified legislators and reporters, who debated and discussed it for weeks. In a series of unreported cases from 1995 until the present, various individuals (presumably all tax payers) have responded in the following way to the fact that I teach two three-credit classes per semester: "Wow. It must be great having to work *like* only six hours per week." And of course, anecdotal evidence would suggest that every university professor hears regularly how lucky he is that he has "off" in the summers.

Such anecdotes suggest how pervasive popular misconceptions are about what academics do on a daily basis. While most of these misconceptions are due simply to ignorance, they also reveal a general distrust of intellectuals, which happens to run very deep in the American psyche. As Nelson and Watt have argued in response to the Washington lawn-mowing fiasco, "what fuels this public rage is a misguided class resentment about the uniquely flexible schedules faculty members enjoy and about the intellectual freedom they flaunt when they take progressive stands on matters of public policy."[8] Fueled by the far Right's hyperbolic and mean-spirited distortions about the so-called Leftism of the modern university, such resentment has escalated in recent years from a mere annoyance to an organized political movement, which threatens both the stability of academic freedom and the schedular flexibility that intellectuals require in order to conduct their research effectively.

One might explain in quite logical terms that the University of Washington faculty member may have been mowing her lawn on Wednesday morning because she was working extremely late in the lab on Tuesday night. One might explain in equally logical terms that in order to teach six hours per week of college-level material, instructors must first research hundreds of pages of material, write lesson plans, hold office hours, and finish grading the forty or so six-page papers turned in last week (to say nothing of the weekly research and service activities of a typical college professor). And to the persons who find it unfair that summers do not include regular teaching loads, one might respond that books and articles—required for merit, tenure, and promotion—do not write themselves. But then one might be labeled "condescending" instead of just lazy.

I bring up these issues here for two reasons in particular: first, few things are more difficult to cope with psychologically as a graduate student than having to explain again and again to family and friends why you are *still* working on your Ph.D. when all of your peers are already lawyers and/or homeowners. You will doubtless encounter, and feel the need to address, any number of ignorant and annoying remarks—from the first day you enter graduate school until the day you retire from academe. Second and more important, given the popular perception of university professors, your own sense of what they do is likely to be as skewed as anyone else's, at least at the beginning of your graduate career. The faster you learn what you've actually gotten yourself into, though, the faster you'll be able to determine whether or not academe is right for you.

Understanding Tenure

At a winter holiday party in 2001, a considerably obnoxious person—who I later learned was my wife's boss—decided that, rather than simply introducing herself, she would ask me the following pointed question: "Oh yes, nice to meet you too. I've been hoping to ask you, why do I, as a taxpayer, have to support your tenure?" Obviously she had been waiting to ask me this question for some time. Rather than explaining that her taxes had nothing to do with the granting of tenure in Connecticut's higher education system (or informing her that I was yet to earn tenure), I asked, "Who *are* you?" Apparently operating at this level of banality was a mistake since it afforded her the opportunity to say what she *really* wanted to say: "I've never understood why you guys have job protection when no one else does." "Maybe you *should* have job protection," I replied. She looked confused.

Of course, I might have responded more thoughtfully by explaining to her the reasons why many academics—after a long period of training and then professional service—eventually are able to earn job security. I might have introduced the concept of "academic freedom" that tenure is designed to protect. I might also have explained that as a protector of academic freedom, tenure is one of the fundamental components of our democracy—that, without it, students at schools like Williams College or Berkeley or Nassau Community College would very likely still be learning about creationism rather than evolution. Instead, I assumed this was all too abstract for the average, outraged Bill O'Reilly fan to bear, and so I simply dropped the conversation. I comforted myself knowing that on college campuses, at least, people understood and revered these basic principles.

I was quite disturbed, therefore, to overhear one of my MA students talking a few weeks later about why she thinks tenure is such an important right of academics: "After all, who else has to work so hard just to obtain a job? It needs to be protected." I say that I felt "disturbed" because tenure so clearly represented in the mind of this future academic a personal reward rather than a gift to the thousands of students she would be privileged enough to teach over her 30 or 40 year career. Now it would be overly cynical of me to assume that she might be failing to do her job in the classroom—either as a student or as a teacher—because of what she had said. And yet I couldn't help but think to myself that her priorities seemed backwards, to say the least.

When academics themselves lose sight of why tenure is necessary, it's time to start worrying. As conveyed in the "Statement of Principles on Academic Freedom and Tenure" (1940), "tenure" is defined by the American Association of University Professors (AAUP) and the Association of American Colleges and Universities (AACU) as follows:

> Tenure is a means to certain ends; specifically: (1) freedom of teaching and research and of extramural activities, and (2) a sufficient degree of economic security to make the profession attractive to men and women of ability. Freedom and economic security, hence, tenure, are indispensable to the success of an institution in fulfilling its obligations to its students and to society.[9]

Tenure is not a sinecure. Nor is it a personal reward. As the primary legal guarantor and protector of academic freedom, tenure is a fundamental necessity in any democratic society (I recommend that you read the entire statement on the AAUP's website). Due to a variety of complex economic and political factors—some of which I have already discussed in this book (see pp. 6–8)—tenure very well may be an endangered species. As a future academic, you should feel a certain responsibility to learn more about the concept of tenure, to study its legal and political histories, and to resist the increasingly vocal and well-organized forces that would try to destroy it. More effectively than I did in response to my wife's boss, you might begin by educating those individuals who would distort tenure by referring to it as either a luxury or a societal burden.

Understanding the Big Three: Teaching, Research, Service

All evaluations of a faculty member's effectiveness take into consideration the individual's contributions in the areas of teaching, research, and service. All three contributions are important in the vast majority of college and university settings, whether a massive Research university or a small liberal arts college. In a scientific survey of "What Search Committees Want," the single most desirable quality sought in a candidate was "Potential for making a positive contribution to the institution *as a whole*," which received a 5.36 ranking on a scale of one to six.[10] Whereas a research university obviously emphasizes research first and foremost, liberal arts colleges tend to place more emphasis on teaching and service. As the survey would suggest, you should be wary, though, of the myths that teaching is unimportant in a research setting and that research is unimportant in a liberal arts setting. Neither is accurate. Assistant professors at high-powered research institutions are often informed that their tenure cases will be evaluated according to a weighted scale that looks something like the following: research 60 percent, teaching 30 percent, service 10 percent. But if one of these professors were to demonstrate utter incompetence in the classroom, no amount of research would be likely to earn him tenure.

As a student in a graduate-degree-granting university, you can bet that your professors face considerable pressure to publish on a regular basis. But you should also remember that only about 10 percent of universities are classified by the Carnegie Foundation as research universities. As the job market looms nearer and nearer, the trick will be to keep straight the fact that your research-university training—with its primary emphasis on publishing—may not be wholly applicable in all job settings. That is, once hired you may be asked to conduct your business very differently than do your own professors, from whom you've learned how to conduct your business. Although this contradiction has led many critics of the academy to complain bitterly of the undue emphasis placed on research by most Ph.D.-granting institutions, such persons should be reminded that the Doctorate of Philosophy is a research degree—not a teacher-training certificate. As we move on to discuss the workload of a "typical" professor, though, you should keep in mind the fact that our model applies mainly to scenarios in which research is highly valued.

Research

In the university at which you are pursuing your graduate degree, research is likely to be very important. At most research universities, the unwritten rule of publishing suggests that professors should publish at least one peer-reviewed article per year and a monograph every 6 or 7 years (i.e., roughly the same amount of time it takes an assistant professor to earn tenure). In practice, certain professors publish much more than this and most others much less, so the average is approximately realized. Since publishing norms vary from phase to phase of one's academic career, though, it will be useful to dissect the issue in more specific ways.

Prior to tenure, professors experience the greatest amount of pressure to publish, simply because their livelihoods and futures depend on their ability to do so. Current practices in most competitive research universities suggest that assistant professors should aim to publish a book or the equivalent in articles prior to tenure review. Typically the first book will be a significantly revised version of the professor's doctoral dissertation. Once hired, the assistant professor should begin immediately to perform necessary revisions since it may take years to find a publisher and see the book into print, as I discuss in chapter 10. The pressure on assistant professors to publish a book may lighten sometime in the near future. Within the last few years, the academic book publishing industry has suffered major setbacks (see pp. 216–17), which is leading a few departments to substitute a multiple-article requirement in place of the book requirement. Defining what exactly constitutes "the equivalent [of a book] in articles" is unclear, and universities like it this way, since it gives them the right to deny tenure to a candidate should his case be problematic in any way. Most books contain five or six article-length chapters, but few published book authors would consider the placement of five or six journal articles the equivalent of publishing a book since the latter requires greater depth and expertise and also demands more respect in academic circles. Think eight or nine articles, therefore, when trying to figure out what constitutes the equivalent of a book.

An assistant professor hired by a liberal arts college may be given similarly vague advice about how much to publish prior to the tenure review. (The reason for such vagueness is the school's desire to avoid litigation should the candidate be denied tenure for reasons other than failing to meet formal requirements.) He may be told that in addition to excellent teaching evaluations and considerable service duties, he is expected to publish "regularly." Of course, at prestigious liberal arts colleges like Williams or Swarthmore, one might be expected to publish as much as professors at research universities. If you are hired either by a less intensive research university or a liberal arts college, however, you should find out right away what constitutes "regular" publishing according to the majority of your colleagues: does this mean that you should publish an article every few years, or will you be expected to publish something every year? Are books ever expected of you?

Several other criteria are used to mark the research productivity of assistant professors. First, departments will assess the professor's general level of activity in the field of expertise. Does the candidate conference regularly? Does he participate in important field-related activities or submit grant proposals for research time and funding? Second, departments will consider the reputation of the candidate amongst his peers by looking at how often he is cited in humanities indexes, and how often he has been asked to review books, evaluate books or articles for publication by journals and presses, or give talks at other universities. And of course, most departments require that tenure candidates submit complete dossiers to external reviewers who are asked to evaluate the file and to testify that the candidate is qualified for permanent employment in academe.

Associate professors must continue to be active and must publish either a second book or the equivalent of this in articles in order to be promoted. Especially important in the review of an associate professor's file are signs of a national reputation. As the dean of my college recently stated, in considering whether to promote someone to full professor, "We are looking to know whether the candidate has achieved the highest level of distinction in his respective field." Full professors can choose not to publish, of course, but most actually publish more than at any other point of their careers since no research means no raises in many schools. While many full professors choose to remain very active in departmental and university service, they have at least the ability to focus more of their time on research and less on service and teaching preparations.

Teaching

For *researchers*, the most sought after teaching load is a 2–2 (meaning two courses per semester). One or more of these four annual courses may be a graduate course. The basic rule is that each college-level class demands about 20 hours of work per week, including in-class time, office hours, preparation, and grading. As I discuss in chapter 6, however, teaching may require many more hours than the daily-recommended allowance would suggest. Certainly, teaching graduate students requires more than 20 hours per week. A typical teaching load at a less competitive research university and many liberal arts colleges is a 3–3. The 3–3 load implies that *some* time is to be spent conducting research. At more elite liberal arts colleges, a typical load is a 3–2, which obviously suggests a university's greater research expectations. In nearly all community colleges and at many mid-level state colleges, though, the teaching load is a 4–4, which leaves little to no time for faculty members to conduct research. Additional factors that influence a particular teaching load's degree of difficulty include the size of the average class and the number of separate courses one teaches each semester. For example, a 4–4 load may be more tolerable if a professor is asked to teach two sections of Shakespeare and two sections of the Renaissance survey, each with 25 students. Although this professor would still spend as many as twelve hours of in-class time, his grading duties might not differ all that much from a professor teaching two different courses of 50 students each. In any case, 40 weekly hours is about the bare minimum of time that dedicated teachers will work.

All universities have mechanisms in place for evaluating the effectiveness of their teachers. In the humanities, good teaching is highly valued since much of the funding for humanities programs stems directly from curious undergraduate students enrolling in classes rather than corporate and government grants, as in engineering or the sciences. In addition to observation days—in which a colleague or administrator observes the teaching of a faculty member—professors are evaluated by their students on both written and computerized forms. An assistant professor might very well be denied tenure because of bad or incompetent teaching. In other situations, merit and/or promotions may be declined for incompetence in the classroom. The expectation of all higher educational institutions is not only that professors will spend time on their teaching but that they will teach well.

Service

"Service" or "committee work" is demanded of just about every assistant and associate professor, and it can take up a tremendous amount of one's time. While most professors are less enthusiastic about their service accomplishments than their teaching and research, they understand the importance of the work and plow through it rather selflessly. The major committees in any department include the Executive Committee, Graduate Executive Committee, Courses and Curriculum Committee, and the Tenure and Promotion Committee (see chapter 11). But numerous other committees abound in humanities departments. In addition, most faculty members will serve at various points in their careers on any number of college- or university-level committees, such as the Graduate Faculty Council or the Faculty Senate. In my first 3 years as an assistant professor at a research university—where commentators often claim there are relatively few service obligations—I averaged between six and ten hours of committee work each week. Of course, administrators such as the associate head or the director of graduate studies will be expected to spend many more hours on committee-related work, though such persons usually are awarded a "teaching reduction" in order to perform their duties. Especially when you are an assistant or an associate professor, you can expect to spend a considerable amount of your on-campus time in the committee room.

What begins to emerge once all the relevant factors are taken into consideration is a very different picture than that painted by most critics of the university. What everyone in the academy already knows, of course, and what should be obvious to anyone who actually looks carefully enough into the matter, is that most professors aren't very lazy at all; in fact, indications suggest that they work many more hours than the average American. Most recent studies, including one by the U.S. Department of Education, suggest that university professors work *on average* about 55 hours per week; I would place the number much higher (perhaps 65 hours per week or higher) for younger scholars, especially those still seeking tenure.[11]

Many people outside the academy—especially those who happen to be miserable in their own jobs—have a difficult time swallowing the idea that someone else might find the work that they do pleasurable or stimulating. Their envy or disbelief leads quickly to resentment and, eventually, to mischaracterization and even to personal insults: "It must be pretty nice to work five hours a week," they will snort. "It must be really nice to have off in the summers," they will snidely remark. "It must be nice to read books all day," they will spit. You can allow such people to annoy you, or you can simply ignore them. Or perhaps you will choose to reeducate them. Or if you simply want to have some fun, try humoring them: "You know, you're exactly right. It's really great being able to read and sleep all day, especially in the summer when I don't have anything else to do." But whatever you do, don't allow them to disrupt your sense of purpose. Focus on your work, and let the record speak for itself. And by all means, mow your lawn whenever the hell you like.

THE POLITICS OF ACADEMIC LIFE

> [A] grad student . . . wrote a letter to the college newspaper, following up some article, and said some highly critical things about the MFA program. These criticisms may have been completely justified, but here are some of the non-visible responses. One professor saw the letter and asked me who this cretin was who had written it. He shook his head and said that in departments in many schools, that person would never get a Ph.D. The head here at the time did not operate in that vindictive fashion, but "seeing that B doesn't get a degree" does not involve any serious unfairness, more a matter of doing no favors. No reminders about this or that deadline. No summer teaching. No fellowships.
>
> Hume, 2004[12]

Just about every young academic entering a new institution, whether as a professor or a graduate student, thinks he can change things for the better. If not, the person probably isn't worth the opportunity he has been granted; put simply, people want new colleagues who can contribute immediately. At the same time, considerable danger accompanies the desire to affect change, and one must learn very quickly how many toes are out there to be stepped on. Institutions as hierarchical as academe are politically complex and can be rather frightening places, and so caution is always advisable. The trick is figuring out how to perform your job effectively and contribute to the life of a department without offending everyone around you. Other than sheer insensitivity or a complete lack of political savvy, ambition and success tend to offend people more than just about anything.

Of course, well-intentioned ambition is very different from careerism or the sort of misguided anger demonstrated by the MFA student in Hume's anecdote. The latter will simply get you into trouble, whether or not this should in fact be the case. In your daily affairs, common sense, sincerity, and reason will protect you more than any armor can. But you should also be sure that you have all of the facts straight before speaking up or committing your good intentions to action. One thing that I *continue* to discover as an assistant professor is just how strong institutional memories tend to be in academic departments. That is, I am continually reminded of how little I know about the reasons why so many things are the way they are in my department. There usually are reasons, though, and they have inevitably to do with past trials and experiments—both failures *and* triumphs—which, in some cases, no one remaining in the department is old enough to remember. Above all else, you must show respect for the ghosts that linger in your department.

Let's push Hume's anecdote a bit further for a moment. Let's just say that this particular MFA student happened to be complaining about the fact that the department's creative writing program has no separate budget for supporting such things as guest lectures from prominent writers. Because the MFA program has to rely on the Department of English, only about $7,000 per year is earmarked for visiting writers. What the student doesn't know, though, is that the program *had* operated with a separate budget about 10 years earlier. When the current dean of the college took over the reins all those years ago, however, he decided that the creative writing program was a waste of money, and so he cut the speaker fund to $2,000. In an effort to save the program, the Department of English absorbed it; because the English Department was able to generate some more money by adding more majors and, at the same time, limit new expenses by relying on its existing resources, the head was able to come up with a few thousand extra dollars for the lecture series. The last thing the creative writing program director wants, of course, is one of his students badmouthing the English Department, since the very existence of his program depends upon the generosity and fairness of the department head of English.

You'd be surprised by how often this sort of thing occurs in academe. While the MFA student may have meant well in seeking more autonomy for his program, he was ignorant about its past and incapable of understanding the potential consequences—both personal and institutional—of his actions. By talking to people in the department, he could have learned a thing or two about the reasons for the problem, could have discovered how to work within the system rather than against it, and could even have helped to change it for the better. Instead, he managed to make things potentially worse and alienated his professors in the process.

The most important advice I can give you about departmental politics is to appreciate the long-term value of short-term observations and experience. Make sure you know which game you're playing before you deal the cards. Especially early on in your graduate career, make it your personal policy to sit back and try to understand the way things work. Read the department and familiarize yourself with its back-story. You will discover not only that you are surrounded by a very colorful cast of characters, but also that there are effective ways to work with most of them, however flawed they appear to be. While every department is filled with its own unique personalities, stock characters are no less common here than in other professions. Though you'll be on your own for the most part, you can make it a point to identify immediately and learn to respond appropriately to the following, ubiquitous persons:

The High Priests and Priestesses: Most departments have one or two experienced professors whom everyone respects—the liberals and conservatives, traditionalists and progressives, men and women, the young and old. Find out who they are and get to know them. Watch them carefully. Their example will teach you much about how to act in and out of the classroom.

Deadwood: Faculty members, usually older, who demonstrate a profound ability to contribute almost nothing in any one of the three areas of teaching, research, or service.

The Black Sheep: Typically a single member of a department who has managed to annoy every other member. Usually the black sheep will be the first person to approach you when you arrive on campus, usually in an attempt to influence your understanding of the department and the people in it. The black sheep may be very persuasive, but you'll recognize the paranoid tinge in everything that leaves his mouth. Fortunately, you will know that you have correctly identified the black sheep of the department when two or more persons say to you about the same person, "Hey, that's the black sheep of the department" (other epithets are sometimes used).

The Careerists: Those individuals who have stopped (or who never began) teaching and researching for any reason other than to add another line to their CVs. You will be able to identify them by their lack of sincerity, their refusal to serve on any committees, and their pseudo-intellectualism. Often they mask their anxieties under a veneer of unparalleled confidence. They are dangerous. Keep your distance.

Service Slaves: The true enemies (and victims) of the careerists, these individuals are rarely spotted outside a department's committee rooms. They often appear to be disoriented. Because early in their careers, they performed their duties competently, the department has never allowed them to escape working on the most important and time-consuming committees. Only rarely are service slaves granted the time to do research.

The Curmudgeons: Though outraged by every change and development since 1968, these individuals suffer from the little-understood inability to retire from academic life. Like the unfortunate souls in Dante's inferno, they apparently have no choice but to endure—*almost* eternally—the sort of existence they find most abhorrent.

The Young Turks: Usually, though not always, recently hired assistant professors, these individuals understand that *everything* is problematic but also believe that they have discovered the solution for every single problem. Though I am very fond of the young turks, I would urge you to separate reality from adrenaline while in their presence.

The Grad Student or Faculty Hall-Talker: Somewhat mislabeled because hall-talkers are sometimes spotted in lunchrooms and mailrooms, these individuals are always available to discuss any variety of interesting subjects and items of gossip. Often they are angry, sometimes merely confused, but they should never be listened to. Graduate students seen hanging around the hall-talkers have been reported to experience problems finishing their own work.

Theory Boy or Girl: Almost always a graduate student, any individual who has determined absolutely that "theory" should be routinely misunderstood, distorted and then adopted as part of a personal conduct code. Theory boys and girls are rarely seen wearing the same outfit twice, and often they are overheard explaining to forlorn-looking individuals that relationships are useless since desire and affection are social constructs. You may choose to chat with one or two of these persons from time to time, but you should never date them.

Life-Long "Learners": Graduate students who have been graduate students longer than anyone else in the department can remember. Ironically, young students regularly seek the advice of the life-long learners because of their vast experience in the department. Unfortunately, life-long learners rarely can offer advice about anything other than how to join their ranks.

Everyman and Everywoman: Perhaps these terms are inappropriate for such admirable individuals, but they refer to the vast majority of the people with whom you will work: those who manage to do their jobs effectively and without complaining.

Considering that these are merely a few of the different characters you will encounter in graduate school, how can you possibly be expected to avoid stepping on people's toes (like I've just done)? Even more problematic, how can you avoid guilt by association if your department is more or less factionalized or if one of these problematic characters should approach or even befriend you? Whatever you do, don't panic. In most cases, simple awareness of what can go wrong and some basic common sense will help you to avoid trouble until you can figure things out for yourself. In addition to accumulating some good old-fashioned experience, though, you might take a few proactive measures to learn more about how academic departments operate:

1. Begin reading regularly, early on in your career, *The Chronicle of Higher Education*. Though at its worst, the *Chronicle* is very expensive gossip, at its best it remains the most comprehensive and balanced account of the economic, political, and cultural issues confronting higher education today. Spending an hour or so each week with the *Chronicle* will help you to improve your academic vocabulary, stay informed about current events and trends in academe, and build the sort of confidence necessary to participate effectively in your own department. Recently the "Careers" section of the *Chronicle* has begun to publish useful advice for younger faculty and graduate students, usually in the form of personal narratives; you may benefit from paying special attention to them. A newer, equally useful source of information for academics is InsideHigherEd.com.
2. Attend as many department talks and other functions as time constraints will reasonably allow. Quite simply, you are likely to learn as much about faculty relations and the political dynamics of your department from a few dissertation defenses and colloquia as from all of the seminars you will take as a first-year graduate student. In addition, you will demonstrate to faculty members and colleagues that you are interested in the intellectual and cultural life of your department.
3. Become a departmental historian. In other words, talk to the secretaries and informed members of the faculty about why the department operates as it does. A sincere interest in departmental

matters should place you in situations where you might learn a good deal from more experienced persons. Ask people what brought them to the university. Ask them about their own career paths. Seek their advice on a variety of issues. Recognize your own inexperience, show respect for those who know more than you do, and you will empower yourself eventually to speak and act with the respect of those around you.

Remember, most of all, that word spreads fast in your department. Seek to learn and keep up with the never-ending departmental story, but make sure you don't become one of its stock characters.

CONCLUSION: THE INTENSITY OF GRADUATE STUDY

One thing I came to believe early on in graduate school was that if my professors were working 65 hour weeks, I should probably be working 70 hours. If they were at the library on a Saturday morning, I should be willing to work on weekends as well. If they used their spring breaks to revise essays for publication, I shouldn't be planning a trip to Key West. Whereas one might deduce erroneously that "because I'm not yet a professor, I shouldn't have to work like one," I approached the issue of work intensity from a slightly different perspective: "Because I will have to write a dissertation, publish, teach well, and serve my department just in order to get a job interview, and since I still have to learn how to do all of those things, I should be working at least as hard as those who already know how to do them." As a graduate student—especially one employed by the university to teach—you will always feel transitional, a hybrid between what you were, an undergrad, and what you hope to be, a professor. Act like your professors, not like your students.

Now that I'm on the other side of the graduate school divide, I believe even more firmly in the logic of this reasoning. I have learned from experience as a graduate faculty member and director of graduate studies that the real difference between the best graduate students (i.e., those who will be competitive on the job market) and every one else has far more to do with excellent work habits and organization than native intelligence or imagination. A surprising number of graduate students are simply disorganized; many others are rather lazy, refusing to do any more than the bare minimum of work required of them. While chapter 2 deals with how to structure your graduate career and chapter 3 focuses on basic organization and time management strategies, it's worth saying a few brief words here about the sort of intensity that will be expected of you as a Ph.D. candidate and a future academic.

I'll begin by offering you my honest advice: if you ever find yourself experiencing more than a *momentary* desire to "get by" in graduate school, do everyone a favor and quit while you're ahead. I offer this advice to be compassionate, not mean-spirited. The fact is that "getting by" will never be acceptable in academe—not at the MA level, and especially not at the Ph.D. or assistant professor levels. In order to succeed as a teacher and researcher in a fiercely competitive profession, you will always need to put in the extra time, work harder than your peers, and push yourself further than you think you can go. The Ph.D. is the mental equivalent of a marathon, and the fact is that no one can simply go out and run 26.2 miles without dropping a lot of blood, sweat, and tears. Of course, this advice will only seem terrifying if you happen to be in graduate school for the wrong reasons. Most individuals who decide to pursue a Ph.D. do so because they are driven to learn as much about a subject as they possibly can. They express an insatiable curiosity to know not only about the specific subject matter they have chosen to study but also about their world in general. And they are willing to spend at least half a decade beyond college to get where they are going. Why else would a reasonable person subject himself to the challenges of graduate school in the humanities? The money? Not very good, actually. The ease? There isn't a more difficult job to obtain. The stress-free, easygoing lifestyle? Obviously not.

To most of those dedicated people who have succeeded in earning a Ph.D. and joining the faculty ranks, the idea of cutting corners is and never has been an option. They work as hard as they can because their work is inseparable from what they most love to do. Obviously, it is not easy to earn a Ph.D. Nor is it easy to balance the demands of an academic life. Few careers present the challenges, pleasures, and personal rewards that professing at a university can offer. You should become a professor because you are completely *obsessed* with your subject and the skills it demands and because you believe it is the single most important thing you can pass on to other people. Nothing else will do.

CHAPTER 2

THE STRUCTURE OF YOUR GRADUATE CAREER: AN IDEAL PLAN

In reference to the indefensibly low retention rates of most graduate departments in the humanities, Cary Nelson has demonstrated that "Graduate students who leave [academe] often report a poor understanding of the structure and process of graduate education."[1] Indeed, considering the numerous, vaguely defined hoops through which graduate students must jump to obtain a Ph.D., it's not difficult to understand why many feel lost or why their lack of direction leads to more serious problems. This chapter seeks to empower you by giving you a very clear sense of direction.

Graduate curricula vary from university to university and across disciplines, and individual programs tend to be highly idiosyncratic. Whereas many of the more traditional programs are rigidly structured, demanding the completion of numerous formal requirements at practically every stage, others operate according to a laissez faire dictum, allowing students the maximum amount of flexibility to pursue their own interests. Neither system is ideal, of course, and both have particular strengths and weaknesses, depending on the individual student in question. The vast majority of programs, however, fall somewhere in between the two extremes; indeed, all things considered, humanities Ph.D. students in North America travel remarkably similar paths on their way to the degree.

In this chapter, I discuss common requirements of typical graduate programs in the humanities and sketch out an ideal plan for completing the Ph.D. in a timely fashion. Especially since course-work, exams, and dissertations are treated in separate, more detailed chapters, I focus here on describing each stage and highlighting the sequence in which requirements are usually undertaken. In order to offer you a helpful long view of your graduate career, this chapter focuses on:

- The "Introduction to Graduate Studies" course
- The MA examination or thesis

- The foreign language requirement
- The process of constructing an advisory committee
- The Ph.D. examinations
- The dissertation prospectus
- The dissertation
- The dissertation defense
- Time to completion: an ideal plan

Since only a few of these requirements may be relevant to your particular program, I have arranged each section so that you may skip around without missing much important information.

INTRODUCTION TO GRADUATE STUDIES OR BOOT CAMP FOR HUMANISTS

Many graduate programs in the humanities require that students pass the equivalent of an "Introduction to Graduate Studies" course in their first semester. Such courses tend to be notorious amongst MA and Ph.D. students, both current and past, but be careful not to buy into *all* of the hype. While it's true that introductory courses often were (and still can be) quite intimidating, especially prior to the ascendancy of critical pedagogy in the 1980s and '90s, they have always been defined by their practical approach to graduate studies in a particular field. Because often they constitute the only specific required course in a graduate curriculum, these courses accumulate over time something like the aura of an undergraduate freshmen composition class. Not only do they serve similarly to "shock" students into the complexities of work at this high level, but as sources of shared suffering, they achieve immediate legendary status for every class of incoming students. The key to contending with such a requirement is in using the highly practical emphasis of the class as a counterbalance to the stress and anxiety it may temporarily cause you.

When in 1995 I enrolled in "English 501" as a first-year graduate student at Penn State, the course was undergoing a major transition from boot camp status—the course's former professors having been accused of every crime except murder—to its current function as a practical introduction to literary theory and research methods. Even then, however, the most intimidating assignments also served very practical purposes. On the first day of classes, for instance, my professor walked into the room, wrote an obscure quotation on the blackboard, explained that it was about Mark Twain, and told us to turn in a twenty-page paper on the source and context of the quotation by the class's next meeting. This was a time, remember, when search engines like Google had not yet been invented and when even e-mail was still new to most of us. Like everyone else in my class, I panicked, ran to the library immediately after class and, figuratively speaking, did not leave for seven days. At the end of the week, not only did all of my classmates turn in the paper, but every one of them had become familiar, even comfortable, with the library's on-line catalogue, the arrangement of the stacks, the microfilm collections, and the helpful librarians. Solidarity had been built amongst the students, and we all acquired in one week's time library skills we could apply in all of our classes, not to mention the confidence to put them to use.

The point is that what may seem most tedious and difficult, even terrifying, about these introductory courses is also what makes them most useful; the Introduction to Graduate Studies course is often the only opportunity for graduate students to study their discipline on the meta-professional level and perhaps the only class designed to train students in the tools and methods they will need throughout their graduate careers. Whether the introductory course in place at your institution happens to be run as a boot camp, a regular seminar, or a friendly forum, never underestimate its importance at the beginning of your career. The course not only will provide you with invaluable skills and information, but it also will help you to establish the work habits you'll employ throughout your lifetime and, nearly as important, it will serve as the basis for how others in your program—both professors and colleagues—perceive you and your scholarly potential. Recognize that you are lucky to be in a program that offers such a course, and don't pass up the opportunity to make the most of it.

THE MA EXAMINATION OR THESIS

MA examinations have become increasingly unpopular over the past 20 years or so. Because most MA exam systems are geared around testing students on a "standard" body of knowledge in a particular discipline, the recent expansions of and attacks on traditional "canons" in all disciplines have called into question the usefulness of privileging certain, or any, texts or cultural artifacts. At the same time, the shift toward professionalism in the humanities has led many departments to require an MA thesis, usually as a substitute for the examination, which is a longer research project that can serve as a practice run for writing the dissertation and/or publishing. In its most extreme form, the examination/thesis debate replicates the more general educational debate about whether content or skills matter most; whereas proponents of the examination system would appear to ally themselves with E. D. Hirsch's promotion of "cultural literacy" or "the network of information that all competent readers possess,"[2] thesis supporters seem more in touch with the sorts of critical pedagogical methods recommended by theorists such as Paulo Freire. Because examinations *and* theses tend to ignore one thing at the expense of another, both systems are problematic. And because the problems tend to be so obvious, both requirements are often so watered down as to lose even their ability to accomplish the limited goals they are set up to achieve. Like many things in life, however, both the examination and the thesis can be useful *if you make them useful.*

Unlike Ph.D. examinations, MA examinations typically test students on standard or fixed lists preapproved by department committees. Generally speaking, knowledge of the approved works serves not only to ensure that students are "masters" of a particular discipline's basic content, but also that they will leave prepared to teach standard courses in their respective fields. Examinations usually are structured by areas or periods, sometimes allowing students to choose three of six areas, and sometimes demanding that they be tested on all six. History MAs at the University of Illinois, for instance, are examined in three fields according to the following guidelines: one field must be geographical (Modern European history, history of China, etc.) or chronological (United States history since 1830); one field must be thematic and comparative ("Race in Latin American and US history"); and one must cover a period prior to 1815. Art history MAs at the University of Virginia, on the other hand, must complete a seven-and-one-half hour examination, administered over two days, consisting of "slide identifications with brief commentary from all fields and three-one hour essay questions in each of a student's two major fields."[3]

Approved examination "material" is more or less vaguely defined depending on the program in question. For example, whereas many literature departments require that students study a list of specific works—an eighteenth-century exam list might include *Tom Jones, Robinson Crusoe, Tristram Shandy,* and so on—others might simply list authors' names: Fielding, Defoe, and Sterne. The latter is usually more demanding since students will have to read multiple works for each author; if Dickens happens to be on the Victorian list, how many 600-page novels should one actually read?

In a sense, such questions suggest the major problem with MA examinations. If graduate students study Dickens in an effort only to discover what they need to know, then the very experience of reading

Dickens, of coming to terms with the complexities of his work on one's own, is likely to be compromised. Examinations are useful only insofar as students are encouraged to use them as opportunities for increasing their knowledge of various subjects. Remember when approaching your examinations that the process of learning the material, of reading and studying, is much more important than your performance on the test itself (although the former is certainly a precondition for the latter [see also pp. 151–52]). Read as much Dickens as you can. Read so much Dickens that, when you finally decide to move on, you feel confident in your understanding of the man's voice, as well as the themes, the historical context, and the aesthetic principles that inform and characterize his work. The only way to ensure that you will do well on a master's examination is truly to *master* the material on which you will be tested.

Generally speaking, theses serve one of three particular functions in MA programs: (1) in departments without MA examinations, as the single formal requirement other than course-work and, in some cases, competence in a foreign language; (2) in departments with MA examinations, as an optional substitute for the examination or, in some cases, as an additional requirement to the examination; (3) in departments with or without the examination, as a three- or four-credit substitute for course-work and/or the examination. At the University of Florida, for example, English MAs are permitted to pursue a thesis option, worth six credits, which serves as a substitute for two required courses and an oral examination on a preapproved list of texts.

The thesis requirement typically entails students working on a sizable research project closely supervised by an advisor or committee. In some departments, the thesis requirement is a mere formality; students simply hand in a revised or expanded seminar paper. In other departments, the thesis must be created from scratch and is scrutinized carefully by experts in the appropriate fields. In programs such as the Art History department at the University of Missouri, the thesis must also be defended orally before a student can receive the MA degree. As with examinations, there are almost as many variations on thesis requirements as there are MA-granting departments.

What the thesis option loses in terms of coverage it gains by cultivating a specialized knowledge of a field and by emphasizing professional development. Students planning to pursue a Ph.D. gain a useful opportunity to practice writing for publication, and extremely mature students might even use the thesis as a testing ground for a dissertation topic or the drafting of a dissertation chapter. Obviously students who plan to leave graduate school with a master's degree may find the experience of having written a 50-page paper on a highly advanced subject less than useful for their teaching.

FOREIGN LANGUAGE REQUIREMENTS

All Ph.D. programs in the humanities require that students possess a reading knowledge of at least one foreign language. Some programs, such as Yale's English department, require as many as three foreign languages. Others require mastery of fewer languages but are more specific about those languages students must learn; the Rutgers Art History department, for instance, requires knowledge of two languages, but one must be German. Since you will be required to select at least one language to study, be sure to consult with your area advisors about what will be most useful for scholarship in your field. Whereas a Renaissance historian would benefit greatly from a reading knowledge of Latin, a literary theorist focused on semiotics would benefit from a knowledge of French. Make a practical and informed decision about which language to study.

Departments evaluate "mastery" in different ways, but graduate students usually can fulfill the language requirement by completing one of the following steps: (1) passing a timed translation test (of a single passage or series of passages), with or without a dictionary; (2) taking one or more advanced (usually literature) courses in the given area; (3) taking two terms of elementary Latin or ancient Greek; (4) completing an "intensive" or "immersion" course in an approved language; (5) passing the TOEFL, if you are an international student. Most programs offer clear guidelines for completing the language requirement so your options will be more limited depending on where you decide to pursue your degree. For example, should you be seeking a Women's Studies Ph.D. at the University of Maryland, you will be required to summarize a ten-page article written in a foreign language and to translate word-for-word one paragraph of that article.

Complete the language requirement as early in your career as possible, and try to do so in the summer months when you will be less distracted by other assignments. The earlier you learn a particular language, the more regularly you will be able to practice using it in seminars and other forums. In any case, be sure to complete the requirement no later than the final summer before your last year of course-work; from this point forward, summers should be used for examination preparations and dissertation writing.

SELECTING YOUR ADVISORY COMMITTEE MEMBERS

Your advisory committee will consist of at least three faculty members—a major advisor and several associates—regardless of where you happen to be pursuing your Ph.D. Some universities require an additional faculty member from outside the student's own discipline. You should find out early on in your career whether or not this is the case at your university, as you will want to enroll in appropriate courses outside your department if it is. Many programs also encourage students to seek out appropriate, often very prestigious advisors from other universities; known as "Special Members," such persons commit to reading the dissertation and presiding at the defense, often in return for a small honorarium and/or a *per diem*. While advisory committees vary, then, in size and makeup, you can assume that your committee will consist of as few as three and as many as six faculty members (there are certain advantages to smaller committees, of course).

Near the end of course-work, and at least six months prior to your Ph.D. examinations, you should select your major advisor. Be sure to consult with this person before appointing associate advisors, since you are almost certainly ignorant about faculty politics and infighting and will want to ensure that lifelong enemies won't seek to work out their personal problems at your dissertation defense. In other words, think about the larger group dynamic in addition to considering those individuals you wish to include on your committee.

While it is common, and even healthy, for students to select one or more advisors who are not experts in the specialized area, the major advisor and *most* of the associate advisors will be active participants in the specific scholarly community you wish to join. Since it is unorthodox for faculty members to serve on committees for students whom they have not taught, it is your responsibility to enroll in seminars with those members of the faculty whom you would like to serve on your committee. I would stress, however, that if a potentially ideal committee member happened to be on leave while you were completing courses, was hired after you completed them, or simply offered a course you could not take for one good reason or another, you should not hesitate to make this person a member of your committee. Often students who complete their MA degrees at places other than their Ph.D.-granting universities find themselves unable in two short years to enroll in courses with every potential advisor. In cases where there is no prior relationship with the potential advisor, you should go to the professor during office hours, explain why you have not worked together before and why this is unfortunate, name your other committee members, and ask whether she would be willing to serve on your committee. The individual may wish to consult with your major advisor or former professors before making a decision, so be kind enough to allow her time and space; in other words, never put anyone on the spot. The last thing you want is a reluctant advisor on your committee.

This last bit of advice raises certain questions about the advisor–advisee relationship *from the faculty member's point of view.* Major advisors, of course, commit a significant amount of time, emotion, and energy to their students. Their duties include, but are not limited to, the following: assisting in the formation of the advisory committee; advising the composition of the examination reading lists; composing and evaluating the examinations; guiding the composition of the prospectus; reading countless drafts of the dissertation; scheduling and running the dissertation defense; preparing the advisee for the job market; writing the student's most important letter of recommendation. In addition, your major advisor will spend countless hours talking to you about your project. Remember that no experienced faculty member will agree simply to take on *anyone* as a major advisee. The relationship with your advisor is something you must earn and work to maintain over the second half of your graduate career.

If associate advisors generally commit less in the way of time and energy, they also receive far less in the way of credit for your successes. Whereas the hard work of major advisors is acknowledged in a variety of ways, secondary advising tends to be taken for granted by promotion, tenure, and merit committees. Good associate advisors agree to serve on committees because they are genuinely interested in a student's work, because they believe they can impart something useful to the student, and because they wish to see the student succeed. The decision about whether or not to serve, therefore, often boils down to rather simple questions: how well do I know this student? How sincerely do I believe in the quality of the student's project? How well do I get along with the student's other advisors? Can I depend on this student to uphold her end of the bargain? How one answers such questions suggests whether serving on a particular committee will be a personally satisfying, even enriching experience, or a frustrating waste of time. If this sounds crass, you should remember that your advisors have to believe in you at least well enough to write you a solid letter of recommendation before you begin your job search. Your goal as an advisee should be first to instill trust in your potential advisors and then to cultivate and enhance it after they have agreed to serve.

Ideally you'll select as your advisors those faculty members who are the most knowledgeable about your topic, most active in the research fields you wish to join, and most able to assist you in launching a research career. You'll also want to consider those potential advisors that are most likely to suit your temperament. Do you need a taskmaster who will uphold regular deadlines? Or do you need a person who will leave you alone to do your research in your own way? While the process of appointing committee members *can be* stressful, you'll likely be surprised, by the time you need to begin making decisions, by the obviousness of the most useful choices. By demonstrating your professionalism, diligence, and overall excellence during the course-work stage of your career, you'll not only empower yourself to build the best possible committee; you'll create a situation where top advisors are actually eager to work with you. This is a goal worth conquering. Most successful faculty members would testify that a good relationship with an advisor can result in tremendous personal and professional benefits long after the dissertation has been written and defended.

PH.D. COMPREHENSIVE EXAMINATIONS

Few formal requirements cause Ph.D. candidates more anxiety and psychological trouble than comprehensive examinations or "comps." The problems that tend to emerge at the exam stage are largely the result of the "freedom" one gains after the course-work ends; individuals who have grown used to the flexible yet highly structured life of the American student suddenly find themselves determining their own work schedule and struggling to uphold their own deadlines. Individuals who take for granted the healthy benefits of their local friendships and social relations—which, more often than not, are linked rather directly to the classroom experience—find themselves at home alone, reading, without any solid connection to the outside world. And for those who manage to escape feelings of aimlessness and isolation, there are still larger philosophical problems to overcome, questions that plague students making the transition to the dissertation stage: can I succeed at anything other than being a student? Can I continue to do this for 2 or 3 more years? Am I really capable of writing a 300-page book?

The "comps" experience need not be so terrible. In fact, if approached in a systematic fashion and understood within the larger context of the graduate career, comps can be one of the most satisfying phases of your academic life. Chapter 7 focuses on how the deliberate imposition of structure onto your suddenly structureless life can transform comps into a positive and productive experience (see pp. 152–54). Then why bring up these issues at all in a chapter on structuring your career? The answer is simple: one of the best ways to counter the potentially destructive effects of the aimlessness that often accompanies comps is to shorten the time you will spend feeling aimless. Graduate departments are filled with Ph.D.s who take way too much time preparing for their exams.

You should seek approval of your reading lists in your final semester of course-work and finish your exams no more than six months after course-work has ended. Assuming that you complete your course-work in the spring semester, you will have all summer to read the works on your examination lists, so that a fall semester examination time is a more than reasonable goal. Such a plan will not only help you to keep the period of potentially destructive isolation relatively short, but it will also assist you in establishing the sort of rhythm and momentum so vital to the composition of a dissertation.

THE DISSERTATION PROSPECTUS

Most Ph.D. programs require a dissertation prospectus, which must be evaluated and approved by the advisory committee and, oftentimes, by an interdisciplinary college or university committee. Sometimes the prospectus must be orally defended before the author's committee and one or two external readers. In any case, only after the prospectus is officially approved by a departmental or university committee does a Ph.D. achieve the coveted status of ABD, having finished "All But Dissertation." While chapter 8, "The Dissertation," offers technical advice about how to write the prospectus (see pp. 173–76), this brief section focuses on the more pragmatic issues of when to write the prospectus, how long to spend on it, and how to use it during the ABD stage.

Make it a point to complete your prospectus no longer than three months after you've finished your comprehensive examinations. Most universities require a document between ten and twenty pages long, excluding the large bibliography of works to be consulted for the dissertation, which can require many more pages; especially if you've used exams to prepare for your dissertation, you should be able to turn over a well-written, thoroughly revised, 15-page document in little time. Remember to take the word "prospectus" seriously. Many students waste a great deal of time attempting to perfect in the prospectus both the dissertation's central thesis and the individual arguments of each chapter. The simple fact is that your ideas and arguments will change—sometimes drastically—during the course of dissertation writing. Think of the prospectus as an announcement of your dissertation topic and a *hypothesis*, as opposed to a conclusion, about what the project will teach readers. If your advisor expects that you should already have the answers to questions you've only begun to articulate and explore, consider switching advisors.

An effective prospectus should serve as a blueprint of sorts for the actual construction of the dissertation. While you are ABD, you will find yourself returning to it for a number of helpful purposes: (1) a good prospectus bibliography doubles as a research-materials checklist. In beginning a new chapter or section, you will need to consult the existing literature on the topic to be analyzed, and the prospectus bibliography provides the best starting point for each new reading phase. Remember also that each prospectus citation finding its way into your dissertation can be simply cut and pasted into the dissertation bibliography, saving you much time and aggravation later on; (2) the prospectus will also remind you of the original questions and goals that sparked your curiosity in the subject, long after you've grown so close to it and lost the ability to see it in even the most remotely objective way. This is not to say that you should force the dissertation material to fit the language of the prospectus but, rather, that you can use the prospectus to measure alterations in your thinking over time, to compare alternative treatments of a particular problem, or simply to remind yourself of your original reasons for engaging a subject, text, or problem; (3) perhaps most obviously, but also most importantly, the prospectus will serve as your most detailed outline. Upon finishing chapter 5, you will be able to turn to the prospectus' brief description of the next chapter in order to ground and remind yourself where you're going next; (4) finally, since the prospectus is your first foray into summarizing in a few pages a several hundred page document, it will serve as a useful model, once your dissertation has been defended, for writing a book prospectus (see pp. 233–37).

In short, a strong dissertation prospectus should grow directly out of your comprehensive examinations and feed directly into your dissertation. To complete the prospectus in a timely fashion, capitalize on the excitement and energy that should accompany the conception and presentation of your first book-length project. Recognize that bigger and better things are ahead of you.

THE DISSERTATION

Chapter 8 deals extensively with the complexities of dissertation writing. Here, I wish to say just a few things about managing time while you are an ABD. If the problem of an unstructured daily life leads to non-productivity during the "comps" phase of many graduate careers, the continuation of the same problem during the dissertation phase is well marked by the rather amazing fact that about 30 percent of ABD candidates fail to complete their Ph.D.s. Think about this for a moment. Of those remarkable students who manage to complete approximately 4 years of course-work beyond the BA, who master one or more foreign languages, who conquer an MA examination or thesis, who suffer through days of difficult testing during comps, and who write a dissertation prospectus, only 70 percent wind up with a degree. Now avoid romanticizing this fact because it has little to do with the difficulty of pursuing a Ph.D. Students who manage to reach ABD status already have demonstrated the ability to handle such difficulties.

Instead you should understand and seek to deal with the inexcusably negligent and exploitative attitudes with which so many departments approach their dissertation students. Anyone who's been around a graduate program in the humanities for any amount of time knows Ph.D. candidates who achieved ABD status far too long ago and who, even after years and years, have failed to make any real progress on their dissertations. Overworked advisors and departmental administrators shrug their shoulders as if confused by what's going on, but even the most interested and energetic of them can only make so much headway against a larger bureaucratic system that relies too heavily on its graduate students for performing the vital functions of the university. The simple message sent by many programs is that so long as graduate students continue to provide such services as teaching, advising, and increasingly, administration, there is no real reason to help them to get out of school. For graduate students who find it difficult to neglect their own students and their service responsibilities, there seems like precious little time for learning how to write a book. In the most serious cases, which are not all that uncommon, they accumulate thousands of dollars of debt and suffer the humiliation of being stranded in graduate school seemingly forever.

While conscientious faculty members and graduate student activists around the country have already begun to address these problems by emphasizing professional development workshops and courses, improving mentoring programs, instituting policies designed to foster the progress of ABD candidates, and even working to unionize graduate students and adjuncts, the problem is not likely to go away any time soon. As a graduate student in today's quasi-corporate university, you need to approach your dissertation in a highly systematic and relatively aggressive way. You can avoid becoming a lifelong graduate student by paying attention to some of the following tips:

1. Most important, try to finish the dissertation in about 2 years. By constructing a calendar or ideal schedule at the prospectus stage and forcing yourself to meet your own strict, self-imposed deadlines, 2 years should be more than enough time for writing an approximately 300-page document. You can help yourself by budgeting for as much free time as possible in the summer months. Seek to write at least one chapter per semester (much easier than writing three seminar papers, right?) and two each summer, and you will be on your way. For more on managing time at the dissertation stage, see pp. 177–82.
2. Next, avoid agreeing to teach more than two new courses at the dissertation stage. Often graduate students find it difficult to say "no" to any new courses, especially advanced topics courses, which they may be offered after they have completed comprehensive examinations. While you should certainly try to diversify your teaching experience as much as possible while in graduate school, try to be reasonable as an ABD. Although your experience of having taught five different courses will undoubtedly impress search committees, such an accomplishment will mean very little if you've taken 5 years to write your dissertation. Limit the amount of time you need to prepare for class by teaching what you know, and you will be a more efficient writer of your dissertation.
3. Finally, begin attending your department's job market training sessions at least 1 year in advance of the time you actually plan to go out on the market (see pp. 258–60). Since you will very likely still be working on your dissertation while job searching, you need to account for the fact that preparing applications and interviewing constitutes a full-time job in itself. The more informed you are about this process, and the more time you allow yourself to prepare for it, the less time you will need to steal away from your dissertation.

While there is no question that most departments should do a better job of preparing students for the dissertation stage and sheltering them from the various distractions that hinder their progress, motivated and diligent students can graduate in a reasonable amount of time by making their dissertation the number one priority in their lives. Be protective of your time; no one else will protect it for you.

THE DISSERTATION DEFENSE

Schedule a defense date prior to the time when you are most likely to be interviewed for jobs. While there are some good reasons to conduct a highly selective job search while still ABD (see pp. 256–58), recognize that your chances of obtaining a desirable job are slim to none without a dissertation in hand. Because the market in the humanities is so brutally competitive, search committees logically see little incentive in hiring individuals who have not yet proven that they can finish their dissertations. By defending prior to the first round of interviews in your discipline, you will make yourself a decidedly more appealing and competitive candidate.

Nine times out of ten, the defense is less of an actual "defense" of the dissertation than it is a public presentation of it, along with an opportunity for committee members and others to ask questions or suggest revisions—either for submitting the dissertation to the graduate school or transforming it into a book. While some departments maintain a certain degree of formality regarding how defenses are conducted, and while some still feature rather harsh interrogations of the candidate and her project, most defenses tend to be rather tame affairs. Since the unwritten rule in most programs is that major faculty advisors are responsible for sending only fully prepared dissertators to defense, most candidates enjoy their defenses as celebrations—the final step in the long process of earning a Ph.D.

TIME TO COMPLETION: AN IDEAL PLAN

As this chapter has suggested, much of the difficulty of obtaining a Ph.D. stems from failure to understand the various requirements for the degree, the manner in which they are related to one another, and the amount of time each should take. As a result, too many graduate students spend an inordinate amount of time in school, accruing in the process significant debts, and often suffering the psychological pangs of diminished confidence and enthusiasm. By focusing early in your career on the particular requirements you must fulfill and by planning for the long term, you will increase your chances of avoiding such a fate.

Figure 2.1 offers a template for time to completion—an ideal schedule for finishing the Ph.D. in less time than the national average of 9 years.

While every individual program features its own unique obstacles to smooth sailing, the six-and-a-half-year plan for completion of the

Year One:	Complete introductory course and approximately four to five other seminars.
Year Two:	Pass MA examination or thesis requirement; apply for admission into Ph.D. program; complete approximately six more seminars.
Year Three:	Begin selecting advisory committee members; fulfill language requirement[s]; present first conference paper; complete approximately six more seminars.
Year Four:	*Fall*: begin composing comprehensive exam lists; have some sense of a dissertation project; complete approximately 3 seminars (seek to draft dissertation chapters through course work assignments). *Spring*: seek approval of exam lists; continue working out dissertation project; complete course work (seek to draft dissertation chapters through course-work assignments); begin reading *The Chronicle of Higher Education* regularly. *Summer*: study for comprehensive exams.
Year Five:	*Late fall*: Complete comprehensive exams. *Winter "Break"*: begin drafting dissertation prospectus. *Spring*: seek approval of dissertation prospectus. *Summer*: begin writing dissertation (aim to draft two chapters); circulate an essay for publication.
Year Six:	Write dissertation (think in terms of two chapters each summer and one during each semester); attend job market training sessions one year early, in fall. Send out for publication at least one (no more than two) dissertation chapters.
Year Seven:	*Fall*: prepare job application packets; complete and defend dissertation prior to interviews. *Spring*: graduate and celebrate.

Figure 2.1 Time to completion: an ideal plan

MA and Ph.D. remains realistic and is applicable across all humanities disciplines and specializations. Six-and-a-half years is, in fact, an ideal space of time for completing a Ph.D. since assistant professors are allotted roughly the same amount of time for proving that they should be tenured. By demonstrating that, as a graduate student, you were able to balance the pressures of research, teaching, and service in roughly the same amount of time it will take you to earn tenure, you offer nothing less than proof to a search committee that you can handle what's ahead.

CHAPTER 3

ORGANIZATION AND TIME MANAGEMENT

In approximately six years of graduate study, your goal will be to learn how to conduct serious research, read hundreds of books and articles, refine your teaching until you can do it well, publish at least two articles, write a book-length study, serve on a number of departmental and/or college committees, and fight for and obtain a tenure-track position—among other things. Feeling stressed out already? If you are human, the answer probably is "yes," and the bad news is that stress is likely to be a fairly regular visitor from now on. This chapter is designed to help you deal with and even begin alleviating the stress caused by the overwhelming demands of professional academic life.

One conviction that I will happily repeat throughout this book is that success in academe has far more to do with organization and diligence than native intelligence. While the last trait might explain why a person has chosen to confront the challenges of graduate school and might even define the limits of how far he can go in his career, it won't get anyone very far on its own. What most often prevents success—other than a lack of passion—are poor time management and inadequate organizational skills, which lead directly to increased stress levels and inefficiency. In order to help you manage better both your time and your work materials, this chapter covers a range of subjects related to organization, such as:

- Establishing a daily schedule
- Prioritizing responsibilities
- "Family planning"
- Making the most of summers and "Breaks"
- Filing and storage

In addition, the chapter discusses the process of creating a solid CV—one of the major tools of organization in academe.

Time Management

Since every person lives his own life, there is no single, correct way to manage time. Therefore, I should like to stress up-front that I view my own personal time-management strategies as instructive only insofar as they represent specific *examples* of general principles worth your consideration. You should remember that the single most important strategy of organization in academic life is long-term planning, as discussed in chapter 2. What follows here is focused on daily activities and short-term planning.

The Benefits of Routine

The biggest leap of maturity I made between the MA and the Ph.D. involved the regularization of my daily schedule, which is not uncommon. As a new graduate student, still single and without roommates for the first time in my life, all I seemed to have on my hands was flexible time. Surely, you know the story: get up every morning at a different time; begin work every day at a different hour; start each day's work with a different task; go to bed each night at a different time; rinse and repeat the next day. While I continue to believe that schedular flexibility is perhaps the greatest practical benefit afforded by the academic lifestyle, I've also come to realize that serious problems can result from too irregular a daily schedule.

Because of the circadian rhythms that so influence our behavioral and physiological functions, our bodies—which, contrary to what Descartes said, affect our minds rather directly—appreciate what they find to be predictable and tend to reject what they find to be unusual. Although every individual's endogenous rhythms are unique, all of us experience more energy and fewer health problems when we function more or less in accordance with our body's natural clock. Especially for writers and people whose work depends primarily on clarity of mind, such energy is crucial. You may be an extremely early riser or you may be a vampire—it doesn't really matter. What does matter is that you establish some sort of daily routine that manages to respond to your body's natural rhythm and, at the same time, trains it to accommodate your professional and personal needs.

Doing so will be very difficult during the first few years in graduate school because your schedule will be determined largely by irregular seminar meeting times, and you still will be learning how to meet deadlines. Nonetheless, your goal should be gradually to regularize your daily schedule so that by the time you begin writing your dissertation, your work habits will be well established. My own daily work schedule, which I trained myself to follow while studying for comprehensive Ph.D. exams, finds me at my desk no later than 7:30 AM and gets me "home" no earlier than 6:00 PM or so. I tend to eat lunch every day at about noon and dinner around 7:00 PM. I go to bed at *roughly* the same time each night. Of course, I will frequently need to work extra hours—when I receive student papers, have to meet an article deadline, or take an hour during the day to visit the doctor—and I am more than happy to stay out late on weekends and certain other nights, but my body can easily adjust to the rhythmic irregularities because the routine is so well-ingrained that it can't be broken. For the most part, an approximately ten-hour day is more than adequate for most academics, especially since we really can *work* for most of this time. What I mean by this is that whereas many office employees waste a tremendous amount of time commuting to and from work, chit-chatting at the water cooler, and lunching for a full hour, we can more easily limit our breaks and avoid distractions, especially when we are able to work at home.

Many academics follow a regular schedule because it allows them to coordinate their activities with those of their spouses and children. Others may begin their workdays at 2:00 PM and work until 11:00 PM, when they get dressed and go out on the town. The point is that no one can prescribe exactly how you should manage your time but, without question, it must be managed.

Prioritizing Your Activities

Although your natural cycle and your own sense of what's important will dictate how you prioritize your research, teaching, and service activities, one thing is for certain: you must be very selfish about protecting your research and writing time. Early on in graduate school, this will mean focusing most of your attention on your seminar responsibilities. Later on, it will mean protecting high energy time for writing your dissertation. But regardless of whether you choose to spend most mornings writing articles or fine-tuning lesson plans, remember that both activities are far more important than your service obligations, which should be fulfilled at times when your energy levels are lower. More mechanical activities such as grading quizzes, checking e-mail, and proofreading might also be saved for those low energy troughs in your daily cycle.

With few exceptions, I begin every morning at my computer. In the summers and over the fall, winter, and spring breaks, I try to stay there all day, since these are the best times for writing and conducting research. During the school year, I may only be able to write for an hour or two each morning before I need to ready myself for teaching, and I expect that most of my day will be focused on teaching- and service-related activities. Since I love to teach and believe in the importance of service, these are activities that I look forward to; at the same time, I realize that I am less able to focus adequately on them if I feel frustrated about my scholarship. The two hours I give over to my research, therefore, are utterly crucial to my sense of having accomplished something important each day, and they allow me to be a more effective teacher and committee member. For this reason, I protect them as I would my baby. During this two-hour period each morning, I refuse to answer the phone, to respond to any knocks on my office door, or to schedule activities of any kind. Only after I have completed my research do I make myself available to the world again.

Since as a graduate student you have a lighter teaching load than most faculty members and far fewer service obligations (and since you don't want to be in graduate school forever), you should try to spend the bulk of your day on research-related activities (i.e., seminar work, examination preparation, or dissertation writing), which will be harder to do at the beginning of your teaching career. Two hours simply will not be enough. But regardless of the amount of time you spend on your research each day, you must learn to be as protective of it as possible. You may be surprised by how difficult such a habit is to develop. Parents and family members must be asked not to bother you during this time. Friends must be prevented from distracting you. Students must be provided with adequate, alternative options for meeting with you. Televisions must be kicked in.

Many people feel that to be protective of their research time is to be negligent as a teacher. Nothing can be further from the truth. Not only will your teaching-related activities be more focused after you have unburdened yourself of the feeling that you're "not getting anything done," but managing your time in one area of your professional life will also help you to do it in other areas as well—especially teaching.

Managing Your Students

Other than preparing lesson plans, which I discuss in chapter 6, the most time-consuming aspects of teaching are grading and conducting office hours. Of the two, grading can cause a great deal more trouble unless the teacher is willing to systematize his personal practices. I can remember very clearly, during my first year as a teacher, spending half an hour or more on each of my twenty-five students' papers. Since I was teaching freshman composition, the students turned in seven papers over the course of the semester, which means that I spent approximately 90 hours that fall just grading papers. Over the next few semesters I struggled to bring down the average time spent on each paper. While to this day I remain willing to spend an unlimited amount of time with papers that require it (I spend about an hour grading every graduate seminar paper), I practice a personal policy of "fifteen minutes per paper," which I would recommend as the ideal for beginning teachers (I explain how to do this in chapter 6). Since papers differ in size, you may wish to work out a plan based on minutes per page—say three to four minutes per page. While my trials that first year were part of an important learning process, one that every teacher needs to experience on his own, one should also have a clear idea about what constitutes an appropriate amount of time for grading student work, and one must seek to realize that ideal. Make no mistake: the quality of a teacher's grading has nothing to do with the number of hours he spends doing it. Rather, it has to do with one's ability to develop an effective system for approaching student work with a sense of energy and purpose, rather than dread.

While it is necessary and highly recommended that you should always be willing to meet students by appointment, you should encourage them to visit you during office hours, and you should never schedule an appointment that disrupts your regular research time. Most professors hold a minimum of three office hours each week—one hour for each hour of in-class time—usually breaking them up into two or three separate meeting times. In reality, many of us spend twice as much time talking with and instructing our students. In determining how to manage office hours, though, keep a few important ideas in mind: first, in order to cut down on the number of meetings by appointment that always require a disruption of your regular or ideal schedule, be sure to arrange one session on the MWF cycle and one on the T/TH cycle. Since many undergraduates will try to set up all of their classes on Tuesdays and Thursdays, they will have no free time to meet with you on those days. By scheduling a regular Monday or Wednesday time, you'll reduce the likelihood of having to meet them at 7:00 PM on Tuesday night. Second, make it clear in your syllabus that you would prefer to meet with students during official hours, "though I am *willing* to meet with you by appointment should you have a scheduling conflict." You should not simply agree to an appointment because a student informs you that "Wednesday morning really doesn't work for me." Is this so because the student will be in class at that particular time or is it because he doesn't have classes on Wednesdays and would prefer not to have to come on to campus? My point is that students should go out of their way to meet with you before you go out of your way to meet with them. Finally, work to keep office discussions focused on course-related issues. Give students direction by asking them what they would like to learn from or accomplish at that day's meeting. As I say again in chapter 6, office hours represent a wonderful opportunity to work closely with students and to get to know them as individuals, but be careful to manage them properly or they will become a terrible burden and a drain on your time.

Eight Days a Week

Much of the time, you won't be able to avoid working on weekends. Even if I could stick perfectly to my daily ten-and-a-half-hour work schedule, I would need to work 5–13 weekend hours just to meet the average number of hours worked by American university professors. What you can do, however, is learn to be reasonable about what you do on weekends and when you do it. For one, you can choose a work time that minimizes disruptions of your personal and family time. Since my wife prefers to sleep in on weekends, I get up early and work until noon, which allows me to spend the rest of my day with her and my son. If on a Thursday I realize that I'll need to read two books and grade ten papers by Monday, I'll tackle the papers on Friday afternoon since I can more easily sneak in reading at various times and places over the weekend—in the living room while my wife reads her own book and my son naps, in the backseat on the way to Aunt Joanie's barbeque, or in the beach chair while I catch some rays. I can update attendance books while watching the Yankees. I can copy edit a manuscript while sitting at the park. Be smart about which work you save for the weekends.

Three Hundred and Sixty Five Days a Year

Holiday breaks and summers are not resting times (though you should try to get *some* rest during them). While it's nice to think that Nature might have granted to academics four months of relaxation withheld from the rest of humanity, it's safe to say that a refusal to work over breaks constitutes an abuse of the system, not a benefit of it. No one expects Christians to work on Christmas day or Jews to work on Yom Kippur, but those holidays last only one day. Breaks are extremely useful times to focus on research since they present relatively few committee and teaching distractions. Stay near your university during the summer (i.e., do not go home [you can never really go home again, anyway]). You will need the library and the stimulation of your friends.

If you are lucky enough to be enrolled in a graduate program that offers summer teaching or research assistance opportunities, seize them. If your graduate program offers no obvious opportunities for on-campus summer employment, complain loudly, since it is in the best interest of research universities to keep their graduate students on campus year round. Then, while the university is busy ignoring your complaints, seek on-campus work in less obvious places. Whatever you do, accept a job at the local donut shop only as a last resort. Working eight hours on donuts—and I mean no offense to the people who do so—will only make you a zombie during the rest of your waking hours. By landing an academic job (especially a teaching or research-related one), you will not only deepen your understanding of the institution and what goes on there; you also will make the transition smoother from work to research. What's likely to be more natural? Moving from your job at the circulation desk of the library to the stacks, or trying to find the connections between a glazed cruller and Jonathan Culler?

Finally, try your hardest to budget your paltry earnings during the year so that you can minimize work hours during the summer. I know this is kind of like saying to a manic depressive "you just need to smile more," but tremendous perks come to those who value their summers as the single most important research time in an academic year.

Family Matters

I'll be honest and admit that I was reluctant at first to include a subsection like this in a book about managing an academic career. I agreed to do so only because several of my graduate students, who read an early draft of the book, implored me to address some of the personal problems that academics deal with but almost never discuss—mainly, the difficulty of balancing an academic career and a fulfilling family life. One of the younger students explained to me that the single biggest headache she has to contend with as a graduate student is her parents' seeming inability to understand why she is "wasting her time" in graduate school or why her "schooling" should prevent her from coming home more often than she does. While this may sound like a trivial matter, the student claims that her constant feelings of guilt and frustration are beginning to take a serious toll. Another student expressed his concerns that becoming a professor would mean never having time to spend with his wife or children. And several women articulated their fear that having children before tenure would prevent them from succeeding as professors. All of these anxieties are perfectly understandable, and it's probably safe to assume that most academics struggle with them at one point or another, especially early on in their careers.

Indeed, unless you come from a long line of academics, and unless you seek to educate your non-immediate family members about academe rather quickly, they are just as likely to add to your stress over the next few years as to alleviate it. You might have them read this book; if they are willing to understand what it is that you do, you may experience relatively few problems. If they show no active interest in learning about your profession, you'll probably spend a good deal of your time over the next few years explaining to your cousin Vinny why you don't really have your summers "off." Now I realize that advising people how to manage time with their family is kind of like declaring which religion they should practice so I promise to be general in what follows; I do agree with my students, though, that it's important to say at least a few words about the matter and to let you decide which advice is worth keeping and which should be thrown away.

Family Demands on Your Time

Since most people's work week ends on Friday afternoon, it may be unfair to blame them for assuming you have nothing better to do on Saturdays and Sundays than attend your second cousin's third child's fourth birthday party. The problem is exacerbated by the fact that people think of you still as a student, not a busy professional. Unless you happen to attend a graduate program extremely close to "home," though, you will quickly fall behind in your work if you agree to attend too many of these extended-family affairs. Further, you will spend time traveling that might be more valuably used for resting or recreation once your work is done. One wonderful benefit of popular misunderstandings of the academic life is that academics can generate any variety of believable excuses—true or not: "I would love to attend little Jake's birthday party, but the professor I'm assisting gave me a Monday deadline that requires weekend work." The "Wish I was there" birthday card is an especially nice and thoughtful extra touch.

Immediate family issues are trickier, of course, in part because you probably want to see your parents and siblings now and then. If this is the case, I recommend some of the following strategies for coping with family demands: (1) Educate them. At the very least, your parents should know what you do on a regular basis, how long it will take you to complete your long-term goals, and where you hope your work will take you in the future. If the people who love you understand even vaguely the nature of your profession and can empathize at all with the pressure you're under, you have a chance of keeping things civil when you decline their various invitations or head out to the library for a few hours during a weekend visit. (2) Learn to compromise. The next time your mom asks you all bleary-eyed whether you're coming home for the weekend, you might consider saying "yes," but then split the weekend in half: "I'm coming home Friday night but, unfortunately, I will have to leave Saturday afternoon so that I can work on Sunday." Since she'll be happy you said "yes," she probably won't mind the attenuation (though, come Saturday afternoon, she may try to convince you to stay over). (3) Create a workspace in your hometown. If the local library or neighborhood Starbucks seems more conducive to grading papers than your mom's kitchen, try setting aside a few hours each day you're home for leaving the house and getting some work done. Tell dad you need to read for three or four hours but that you'll be back in time for lunch. (4) Finally, recommend that family members visit you from time to time. Though most relatives will seem confused when you tell them that you'd prefer not to drive five hours every other weekend, they probably would be unwilling to drive this far, this often themselves. Insist that they do so once in a while. It will be good for them and for you. The time and stress you'll save from not having to travel, and the comfort of being near your university (and its library), will make staying around your new home more than worth your while.

Getting Married in Graduate School

Since the legal institution of marriage has little to do with the strength of a committed relationship and since it doesn't include gay couples outside Massachusetts, I use the term liberally to apply to any live-in partnership. In the university where I received my Ph.D., very few graduate students were married. In the one where I now teach graduate courses, many of the graduate students are married. But in both places, marriage is perceived positively for the students who choose to enter into it. I bring up the subject here only because so many people harbor understandable concerns about the difficulties of balancing work and marriage. Whereas the benefits of being single in graduate school seem quite obvious (they include superior flexibility and, often, fewer distractions), a strong marriage can actually help some students to manage their time. First, marriage cuts out the frustrations and distractions that dating can cause. More important, if academic marriages are to work, they require that each partner regularize their work schedules enough to maximize the amount of time that might be spent together. And as we stated above, a more regular daily schedule can result in greater efficiency and productivity. In the end, marriage is no more or less ideal for academics than for any other professionals. But marriage should under no circumstances be viewed as an obstacle to a successful academic life.

You may want to discuss with your partner, rather early on in your graduate career, what sort of geographical restrictions will eventually be placed upon your job search. I've seen many happy couples go through a very rough time because of their failure to deal with this issue until just prior to a job search. Obviously, you will face more difficult prospects of getting hired if you restrict your search too severely. Doing so may be the right thing for you and your family, of course, but you should at least know the difficulties you're likely to face long before that job list actually comes out.

Having Children in Graduate School

In an ideal world, you'd wait until your dissertation had been revised for publication (or other major tenure requirements were met) before starting a family. Surviving graduate school and especially an assistant professorship will be extraordinarily difficult unless you make them your first priorities. Obviously, if you have a child, he should be your first priority. Again in the ideal world, then, you'd finish your Ph.D., get settled into your new tenure-track job and only then consider beginning a family. In reality, though, there are many reasons why people might decide to have children in graduate school; and since so many people "go back to school" later on in life, many students will enter graduate programs already having had one or more children. So while the ideal scenario has people waiting until later to have children, reality—and biological clocks—can make the waiting very difficult.

Again, in spite of myths about academe's utopian practices and policies, the reigning tenure and promotion system has been unfair to women (and parents, more generally) since its inception. Since thirty-five years and older is widely regarded in the healthcare community as an age for high-risk pregnancies, a serious question arises: when *are* women who wish to have children supposed to do so? If women take several years off between college and graduate school, as men often do, they will very likely push their doctoral studies well into their thirties.

If, on the other hand, women go directly to graduate school, finish on time, and get a job their first year out on the market, they are very likely to be on the tenure track between the ages of about thirty and thirty-five. This dilemma may explain why, despite the fact that women outperform men in practically every academic category, they still only make up about 39 percent of the professoriate.

Fortunately organizations such as the AAUP and the MLA have begun to address the problem by suggesting ethical guidelines for safeguarding women's and parent's rights. For instance, not only is it illegal for members of departments in many fields to ask candidates questions about their families during interviews, but more and more university departments are granting new parents (unpaid) year-long leaves during the probationary tenure period (sometimes referred to as "stopping the [tenure] clock"). Despite such positive improvements, many women continue to report violations of the guidelines during interviews, and new dads who take off time or stop the tenure clock are often ridiculed in various indirect ways by administrators and colleagues alike (as well as family members, in some cases), just as in corporate America. Clearly the problem is far from being solved, and I would not be so pretentious as to claim that the solutions to it are obvious or easy.

If you do have children in graduate school, you can approach your workload in one of two admittedly unappealing ways. You can accept the fact that it might take you longer than your classmates to earn your Ph.D., which in turn might hurt your job prospects. For many graduate-student parents, there is no choice but to accept this difficult fact. Another option is to do what so many dedicated parents and overworked Americans do on a regular basis: make more personal sacrifices and work even harder than before. Since my son's birth, I've discovered exactly what I value and what I am willing to sacrifice. Since I would prefer to spend every waking hour with him but have no intention of giving up my research career or compromising my teaching, a few hours of sleep each night have gone out the window; television viewing, although never a favorite activity of mine, has gone too; my beloved Yankees games have become mere background noise; and ironically—considering my advice throughout this chapter—my usual schedule has regressed at times into a pre-doctoral-like mess of random work times and momentary bursts of productivity. But I always put in my hours, and I appreciate the superior flexibility of my academic schedule. Indeed the best news I can give to graduate students with kids—or those thinking about having kids—is that while extremely hard work and exhaustion are inevitable, the academic life at least offers you multiple opportunities for managing your time in ways that you feel are best both for you and your family.

Exercise and Hobbies

In Xenophon's *Memorabilia*, Socrates declares to Epigenes that there cannot be a sound and healthy mind unless it is housed in a sound and healthy body, an idea still promoted by scientists and healthcare workers alike.[1] Since all work and no play will make you both an unhealthy *and* a dull person, I would emphasize, here, the importance of maintaining some sort of daily physical activity in graduate school. Although you may often feel as though you have no time for exercise, sticking to a daily routine will bring immense personal and professional benefits, including the following: (1) exercise can help you to structure your day. While class, meeting times, and various deadlines are likely to fluctuate from day to day, your daily workout time can serve as the anchor of your schedule, which in turn will help you to organize the rest of your day. Regardless of all other things, the knowledge that you will be on your bicycle at 3:00 PM should force you to think seriously about what you must do before and after this time. (2) Exercise will help you to relieve stress, which in turn, will make you more productive during work hours. I trained regularly for marathons during my dissertation phase mainly because the long runs demanded that I shift my attention away from Renaissance England for a few moments each day. After finishing a run, the idea of returning to my dissertation always seemed easier, even welcome. (3) Staying (or getting) in shape will result in greater confidence, which will of course pervade everything you do. (4) Exercise will help you to stave off various illnesses, which are, of course, both painful and time-consuming. (5) Exercise clears a space in your day and in your mind for deep and focused contemplation. During a run or a yoga session, you may choose to forget completely what happened in class today, or you may choose to think carefully about what you learned from it. Either way, this sort of private meditation serves its purpose. (6) Most athletes recognize that by renewing and deepening one's energy levels, taking time out to exercise paradoxically results in more quality work time. Not only will exercise improve the quality of your sleep cycle, but your mind will also be sharper during waking hours.

On a related note, you should make it a point to pursue hobbies or other activities that make you happy. Spend a little time in your garden each day. Go to your university's basketball games or musical concerts. Buy yourself a new DVD each time you finish a seminar paper. Keep yourself a well-rounded person or the intensity of graduate study will wear on you very quickly.

ORGANIZATION

The materials used by an academic can be as difficult to organize as one's time. Throughout this book I discuss various organization strategies pertinent to the different skills and stages of your graduate career. In what follows here, I would like simply to offer a few bits of advice about how you might effectively arrange your office space. Think of the materials you will collect, the notes you will take, the lesson plans and syllabi you will generate, and the books you will buy over the next few years as parts of your personal information library. Not only will you glean most of what you know from these materials, but you will also find yourself returning to them again and again throughout your graduate and professorial careers. This is why although I hate clutter, I rarely throw out anything related either to my research or my teaching. In order to be efficient later on, it's extremely important to establish a logical system for storing and accessing all of these materials.

Filing Cabinets

If you need to, use your credit card to buy at least one sturdy three- or four-drawer filing cabinet. Many universities have a site where old filing cabinets and desks are sold cheaply. Office supply places in academic towns sometimes have sales on such items, particularly if they have become scratched or damaged. Secondhand office supply places also sell good filing cabinets for a fraction of the new cost. Few investments will be more necessary for what's ahead of you. Buy also at least 50 hanging folders with tabs.

When deciding how to arrange your files, consider starting with more general categories: designate one drawer each for research (i.e., seminar work, etc.), teaching, and "other" (since "service" won't accurately describe all of the materials you'll include therein). You might consider using the fourth drawer for bills, receipts, and other "personal" materials. In any case, you'll begin filling up the cabinet as soon as your first semester in graduate school. Since your research drawer will include copies of seminar syllabi, seminar handouts, photocopied articles and book chapters, and copies of your own writings, you can imagine how quickly you'll need a separate cabinet for filing your research materials. The teaching files should be arranged by each class you teach, but you might consider also filing away in this drawer pedagogical articles and information. The "Other" drawer might include service-related information, annual TA contracts, and library maps and information, among other things. I recommend creating a file entitled "Other People's Stuff," into which you can insert classmates' CVs, examinations, dissertation proposals, teaching portfolios, and so on. All of these materials, which you should begin collecting from willing donors as soon as you begin graduate school, will be invaluable resources later on. Again, the key is figuring out a system that works for you and being willing to expand as you collect more materials. Without question, you will need to purchase a new cabinet once you begin writing your dissertation.

Bookshelves

Encouraging academics to buy bookshelves is a little like telling a starving person to eat something: it's not really necessary. But how you arrange your books will become an increasingly important issue as the size of your library continues to expand. Should you organize those art history books chronologically, nationally, alphabetically by artist or critic, or by movements? If you have an office space on campus, should you divide your books between the two spaces? Which books should remain on campus and which should stay at home? Obviously your answers to these questions will not make or break your career; the real point is that you devise a system that meets your needs. However you do it, be sure to set aside at least one shelf at home, or use your on-campus office space, for books that you've checked out of the library.

Binders

For later ease of access, you might consider binding in a three-ring folder all materials related to a particular class, whether one you've taken or one you've taught. The binders can then be arranged on your bookshelves, freeing up valuable filing cabinet space. You can be more or less anal about how you mark these binders, at the very least marking the title of the course on the binding or at the most, creating a table of contents page for each binder, as in the abbreviated example in figure 3.1.

Tack-Board

As long as you rearrange it every semester, throwing out the stuff you no longer need, a tack-board can be a useful way to post immediately

"Renaissance Political Criticism" Class: Binder 1

CONTENTS

Course Syllabus	Teacher's Name
Background Lecture Notes	
"Renaissance Literary Studies and the Subject"	Louis Montrose
"The New Historicism in Renaissance Studies"	Jean E. Howard
"Political Criticism of Shakespeare"	Walter Cohen
"The New Historicism and its Discontents"	Edward Pechter
Midterm Paper Assignment	
"Are We Being Historical Yet"	Carolyn Porter

Figure 3.1 Sample contents page for course binder

important information. On your board you might include copies of your seminar and teaching schedules, a list of deadlines for upcoming assignments, a calendar, ideas for seminar papers, and so on.

Date Book/Calendar/Desk Blotter

Since so many people are using PDAs and cell phones for the same purpose, my recommendation of a "date book" should be taken loosely. Not only will such a book be crucial, though, for keeping track of appointments with professors, students, and colleagues, but it will also be useful as an additional way to manage your time more effectively. At the beginning of each semester, make it a point to sit down with all of the relevant syllabi in your life—the ones you've generated for your own classes and the ones given to you by your professors—and mark the dates for every upcoming assignment. You might also consider constructing a one-page list of these deadlines, which you can post on your tack-board.

Personal Computer

Of course, your computer may be your most important organizational tool. Since again how you manage your computer is your own business, I have only a few pieces of advice for you on the subject: (1) Keep accessible your syllabi and lesson plan files for every course you teach. Since you will reuse lesson plans in most of your subsequent courses, you'll want to be able to cut and paste material from older files on a regular basis; (2) bookmark important websites and electronic databases such as your university library, *The Chronicle of Higher Education*, Calls for Papers sites, the *OED Online*, and so on; (3) clean up your files every semester or, at least, every year. Be sure to store on disks anything that might prove useful down the road. Just as you need to keep your bookshelves and filing cabinets clean and well organized, you'll want to maximize efficiency on your computer by the same means.

Remember that all of the time you put into organizing materials now will save you a great deal of time later on. As important, by staying organized you will allow yourself to feel like you're in control of what can often seem like an overwhelming amount of material—a crucial prerequisite for minimizing stress and maximizing productivity.

THE CURRICULUM VITAE

You'll hear much about CVs over the coming years. Like so many important documents in academe, though, the CV is rarely discussed or explained in official forums; without any guidance, you'll simply be expected to generate one at certain points in your graduate career. "CV" refers to "Curriculum Vitae" or "Course [i.e., curriculum] of Life," and it serves as perhaps the most useful and widely used summary of your academic achievements. Unlike a resume, which often includes personal information about such things as one's hobbies and personal interests, and always includes information about one's indirectly-related work experiences, a CV records information solely pertinent to one's academic life and accomplishments. Some common uses of the CV include the following: (1) Conference chairs will often ask that persons submitting abstracts or papers for consideration include a CV along with their submission. Yes, you're right, the point is to see whether or not you've done anything in the past that qualifies you for the present assignment. And yes, you're right again that this practice violates the spirit of a pure peer review system, which should judge the merits of a particular piece of scholarship without consideration of who has produced it or how much else the individual has produced. Do not despair. While such practices may make it difficult for you to land papers in the most competitive conferences in your field, most chairs do make an effort to be fair, and many enjoy the opportunity to "discover" and give a chance to bright graduate students. (2) In situations where your work is to be considered for an award of some kind—a teaching award, writing award, or grant—you will almost certainly be asked to submit a CV. The scenario is very similar to the one described in the case of conference calls for papers. (3) You will be asked to provide a CV to any persons interested in describing you or your work. If you're lucky enough to attract the attention of a local journalist, for example, he may ask for a CV before writing his article. Sometimes conference chairs who have already accepted your paper or abstract will use your CV to write their introductory comments about you and your work. The variations on this theme are seemingly infinite. (4) Most important, you will submit a CV to every potential employer you contact when you go on the job market. At no other time in your career will the contents and style of your CV be more important than while seeking your first tenure-track job.

Long before you start worrying about the job market, though, you should create and maintain a CV, since it's an extraordinarily useful tool for keeping track of what you've done as an academic. Also the process of building a CV will teach you as much about what you *have not yet done* as it will about what you *have* done—hopefully motivating you to devise work plans for the future. While this process can be a bit humbling in the beginning of your graduate career, you'll take great pleasure in watching your CV grow over the years.

So what does a CV look like? While we might discuss any number of variations between disciplines and fields of specialization, CVs typically consist of the following sections. Although I've arranged them in order of importance, the arrangement of your own CV will understandably differ depending on how much you've accomplished in each of the areas. Obviously you should not create a subheading for "Publications" unless you have something to list there. In addition to the figures provided below, you can also look at a solid examples of a graduate student CV by turning to the appendix (288–91).

Name and University Affiliation

Make sure that readers know who and where you are before they begin reading the rest of the CV (see figure 3.2).

Your name should be set in a font slightly larger than the regular text of the CV (which should be 12 point) but not so large as to convey your massive ego. Use a clean and clear font such as Times New Roman or Garamond; avoid cutesy ones like Old English (alert to would-be clever medievalists) or unclear ones such as any of the various script fonts. Clarity will be appreciated by and expected from your readers.

Davida Ariche

374 Peasant St, 2nd Floor; Willimantic, CT 06226
(860) 424–0332 davida.ariche@uconn.edu

Figure 3.2 CV entry for personal information

Educational Background

List in descending order from most recent to least recent your various higher education degrees and university affiliations along with appropriate dates (see figure 3.3). If you are still in the process of completing your MA or Ph.D., provide a start to present date (i.e., "2001-Present") or list the likely graduation date, as in the figure. If you graduated *summa* or *magna cum laude*, you might note this fact, but avoid unnecessary fillers in this section; for example, once you've earned an MA, it doesn't really matter what your undergraduate thesis happened to be about.

Dissertation Information

While still an ABD and after you've defended your Ph.D., you should include some information about your dissertation, as in figure 3.4. You need not list every committee member, and you need not include a table of contents. You should include the title of your project, the name of your major advisor, the date (or projected date) of your

Education

Ph.D., English, University of Connecticut, anticipated May 2004
MA, English, University of Connecticut, May 2000
BA, Philosophy, *Magna cum laude*, Western Kentucky University, May 1997

Figure 3.3 CV entry for educational background

Dissertation

"Mediating Colonization: Urban Indians in the Native American Novel"
Director: Donna Hollen

According to the 2000 United States census, over two-thirds of the more than two million Native Americans counted live in urban areas. My dissertation analyzes how United States Native American writers represent this burgeoning but often overlooked population in a Native American literary landscape traditionally dominated by reservation-based narratives. I examine the evolving portrayal of urban Indians mediating cross-cultural identity in novels by D'Arcy McNickle, John Joseph Matthews, N. Scott Momaday, Leslie Silko, Sherman Alexie, and Greg Sarris. Critical and literary exploration of native transcultural experience often tends toward models that privilege traditional indigenous homelands over the perceived alienation and cultural degradation of the city. I argue that the novels in my study offer a multifaceted view of urban Indian identity that complicates dichotomous models of place in identity politics. Even when they declare the city hostile to native identity, these novelists still recognize, mostly through secondary characters, that native identities can be created in urban spaces. Such an ambivalent view of the city, though potentially undermining strictly traditionalist native viewpoints, nonetheless complements contemporary ideas about the flexibility and adaptability of indigenous identity and culture.

Figure 3.4 CV entry for dissertation description

defense, and a *brief* paragraph describing your argument and contribution. Delete this paragraph once you begin your first tenure-track job since you'll want to emphasize publications rather than work you did as a graduate student.

Work Experience

Record here in descending order any academic employment from your most recent to your least recent jobs. You might mention your work as an instructor, a teaching assistant, a research assistant, or an administrator. Specify under "Teaching" the specific courses you've taught and make clear whether or not you were the primary or sole instructor. You should mention the number of sections you have taught and the number of students enrolled in each course. You should not provide any kind of in-depth description of what you did in that course beyond making clear its subject matter (i.e., do not provide a paragraph description of courses taught). A CV should list such information rather than providing text.

Publications

Few things cause more aggravation for readers of CVs than unclear or, worse, manipulative publication information. You must be absolutely clear when writing your publications section that you differentiate book-length projects, article-length pieces, notes, and encyclopedia entries. Figure 3.5 uses subheadings within the publication section in order to mark such differences.

You must also make clear in each entry your role in the publication process, especially if the piece happens to have been collaborative. List all publications first by order of importance (i.e., articles should be listed before notes) and in descending order from most to least recent. List "Forthcoming" and the projected date for works accepted but not yet in print. Only if you've already published one or more pieces should you list "Works in Progress" since such a category means very little unless you've already proven your ability to publish. If you decide to list "Works Currently Under Consideration," do not list the name of the potential publisher.

Since publication listings will always include the title of the piece, the name of the source, the date of publication, and the length of the piece, readers will figure out sooner or later what type of publication you've earned, and even the hint of dishonesty on your CV can result in the dismissal of your application from consideration. However, as

Articles
"Sinners Among Angels, or Family History and the Ethnic Narrator in Arturo Islas's *The Rain God* and *Migrant Souls*." *Lit: Literature Interpretation Theory* 11.1 (Summer 2000): 169–97. Also to be reprinted in *Critical Mappings of Arturo Islas's Narrative Fictions*. Ed. Frederick Luis Aldama. Bilingual Review Press, 2004.

" 'Dear Billy': H.D.'s Letters to William Carlos Williams." *William Carlos Williams Review* 23.2 (Fall 1997): 27–52.

Notes
"A Look at Basic CV Writing." *Notes and Queries* 1,432 (2023): 1–2.

Reviews
Postethnic American Criticism: Magicorealism in Oscar "Zeta" Acosta, Ana Castillo, Julie Dash, Hanif Kureishi and Salman Rushdie by Frederick Luis Aldama. Austin: University of Texas Press, 2003. Forthcoming in *Aztlán* (Fall 2004).

Birchbark House by Louise Erdrich. New York: Hyperion, 1999; *Muskrat Will Be Swimming* by Cheryl Savageau. Illustrated by Robert Hynes. Flagstaff, AZ: Northland, 1996; and *Rain is Not My Indian Name* by Cynthia Leitich Smith. New York: HarperCollins, 2001. *Multi-Ethnic Literatures of the United States (MELUS)* 27.2 (Summer 2002): 246–49.

From the Belly of My Beauty by Esther G. Belin. Sun Tracks Series 38. Tucson: University of Arizona Press, 1999. *Multi-Ethnic Literatures of the United States (MELUS)* 26.3 (Fall 2001): 233–37.

Figure 3.5 CV entries for publication types

long as everything is clear, a strong publication section will almost always be the focal point of your CV.

Conferences

List the title of the paper, the title of the conference, and the place, date, and sponsor of the event (see figure 3.6). Be sure to specify whether you served as a presenter, respondent, or session chair. List all entries in descending order from the most recent to the least recent. You may also choose to list here any in-house colloquia in which you've participated, or you may designate a separate section of the CV for such a purpose. In any case, be clear about the type of "conference" you're recording.

Awards and Honors

Certain prestigious undergraduate awards such as honor society inductions and fellowships are acceptable on a CV. The Dean's list information is irrelevant. However, most of the entries in this section should demonstrate excellence at the graduate level. Record in descending order from most recent any teaching or writing awards,

"Imagining Native Americans Off the Reservation." Organized three panels for the 2003 Northeast Modern Language Association Conference. 7–8 March 2003, Massachusetts Institute of Technology.

"Seattle's Last Stand: Ethnic Urban Geography and Racial Violence in Sherman Alexie's *Indian Killer*." Presented at the Multi-Ethnic Literatures of the United States (MELUS) Conference entitled "Multi-Ethnic Literatures and the Idea of Social Justice." 9–12 March 2000, Tulane University.

Figure 3.6 CV entries for conference papers

fellowships, grants, or honor society inductions you've earned during your career.

Service

Record any administrative or other service activities in order of importance. If all service activities are essentially equal in importance, list them in descending order from most recent to least recent. Be sure to specify your role clearly, and avoid implying that one-day events or activities constituted a full year's work. Such designations as "2004–05," "Summer 2004," and "April 21, 2003" are important indicators of the types of service you've performed.

Memberships

Record any relevant memberships in professional societies and organizations. You should make it a point to join the major organizations in your field and/or discipline by the time you go on the job market.

Languages

Most people record language proficiencies as filler for relatively empty CVs. Unless your language skills are directly relevant to your field (that is, you are a comparative literature student, classics instructor, medievalist, etc.), there is no particular reason to list them.

References

I recommend that all graduate students provide three or four references. The entries should include the names, titles, affiliations, and e-mail addresses or phone numbers of the selected faculty members and/or administrators. If you are on the job market, seek permission from your referees to include their home phone numbers since interviewing season usually happens over the holiday break when most professors stay at home. Once you are hired in your first tenure-track job, remove the references section from your CV.

I can imagine few reasons why a (humanities) graduate student's CV should be longer than four pages. Be sure to avoid filler in all cases, and do not fall prey to the erroneous assumption that length equals quality. The document should be neat and clean, allowing enough white space to highlight symmetry and structure, and it should not be overcrowded. Your goal should be to produce a CV that someone can scan rapidly, logically, and without any confusion.

Make it a point to draft your first CV sometime before the beginning of your second year as a graduate student, and be diligent about updating and revising the document throughout your career. Ask to see other people's CVs and be willing to show your own to as many people as are willing to look it over and provide feedback. Few documents will be more important to your future than a solid CV.

CONCLUSION

Unlike most professionals, academics are left almost solely in charge of their time, their workspaces, and their work materials. For this reason, it's important to develop strong organizational and time management skills. Failing to do so will simply exacerbate the all-too-common feelings of stress and anxiety that paralyze so many inexperienced graduate students. By maximizing your productive work time, keeping your work materials and tools neat and accessible, and tracking clearly your progress and accomplishments, you'll succeed in alleviating such burdens and live a more healthy professional life.

CHAPTER 4

THE GRADUATE SEMINAR

The three- to four-year long process of completing course-work constitutes the most important stage of your graduate experience, since it helps you (1) to determine your area of specialization and your advisory committee; (2) to accumulate knowledge in your area of specialization (and other areas); (3) to discover your dissertation topic, and (4) to develop the basic work habits and professional research techniques that will carry over into your first postgraduate job. Although graduate courses take several different forms (upper- and lower-level courses, workshops, tutorials, pro-seminars, etc.), I use the term "seminar" in this chapter to refer to all graduate-level courses in the humanities, which are characterized by their excellent teacher to student ratio and their focus on highly advanced subjects.

Several important matters pertaining to the graduate seminar—such as managing time, writing research papers, and constructing oral reports—are dealt with in greater detail elsewhere in this book. This chapter seeks to describe fundamental and practical issues related to seminar work, including the following:

- The process of selecting courses
- The responsibilities of seminar participants
- The meaning of grades
- The relationship of the seminar to your dissertation
- The relationship of the seminar to your teaching

By the time you finish this chapter, you should know what to expect from a typical seminar, and hopefully you will be confident and motivated to make the most of every seminar experience.

WHAT IS THE GRADUATE SEMINAR?

A typical seminar is three hours long and meets once a week, though occasionally seminars meet twice a week and last only an hour and a half. Whereas non-humanities programs often require that students spend up to 40 weekly hours in class, humanities programs tend to limit both in-class time and the required number of courses so that students can devote an adequate amount of time to extremely demanding reading and writing assignments. Although assignments vary drastically from course to course, seminars in the humanities often require exhaustive secondary literature reviews of a specific topic, oral presentations of one sort or another, and a substantial research paper (see chapter 5).

Most humanities Ph.D. students will complete four years of coursework, two in pursuit of the master's degree and two more prior to comprehensive examinations. The typical TA program requires that students take three courses per semester; programs operating on the "quarter system" usually require two courses in each of *three* annual semesters, and programs in which teaching opportunities are limited require four to five courses per semester. In any case, a typical Ph.D. in an MA-granting program will complete *approximately* twenty courses over a three- to four-year period beginning upon entrance to graduate school. In programs without the terminal MA (i.e., five-year Ph.D. programs), students may take many fewer courses, though this will not necessarily be the case. Foreign language courses, teacher-training courses, and professional workshop courses are usually taken *in addition to* these two-dozen or so "content" courses. Unfortunately, there is no magic formula for determining how many of these courses should be focused in your area or field of specialization, but you should try, at the very least, to ensure that 25 percent of your total hours consist of courses in your area.

THE PROCESS OF SELECTING APPROPRIATE COURSES

Several simple guidelines will help you to determine those seminars to take and those to avoid like the plague. In what follows, I've broken them down into the most basic do's and don'ts.

1. Do take most available courses in your area of specialization. The most obvious reason to do so, of course, is to increase your expertise in the field, but you should also consider the numerous practical advantages of pursuing such a policy. First, you will benefit immeasurably by getting to know all of the faculty and graduate students who work in your field. Your area professors are not only likely to serve as your dissertation advisors, but they also will be your most immediately valuable colleagues once you leave graduate school. Excellent graduate

advisors often serve as advisors for life, helping you to cope with the stresses of your first job and the pre-tenure experience, introducing you to important people in your field, endorsing you for a variety of professional activities such as book reviewing and conference participation, and writing you crucial letters of recommendation throughout your career. It is crucial that most of these advisors be experts in your own field. Job search committee members, for instance, would consider exceedingly odd a candidate specializing in U.S. military history with letters of recommendation from two early modernists and a nineteenth-century science historian. Further, a member of a job search committee is likely to find quite problematic the absence of a letter from a well-known, relevant person in the student's department. Even worse, should this search committee member happen to inquire by e-mail into the quality of the student's work, only to be told by the relevant faculty member that the student never bothered to enroll in one of her courses, the committee might very well drop the candidate from consideration.

Strong relationships with other students in your area will help you in countless ways. More experienced students can help you to understand the idiosyncrasies, expectations, and strengths and weaknesses of certain faculty members in your field. Or they might serve as mentors who will give you invaluable advice about such matters as comprehensive exams, dissertation writing, and job hunting. Students at your own level can help you to deal with the stress and anxieties that often affect graduate students, offering friendship, moral support, and healthy competition when they are most needed. Remember also that your classmates today are your colleagues tomorrow—"connections" in your field with whom you might eventually pursue collaborative research, organize conference panels, or simply discuss the matters most important to you.

Finally, taking multiple seminars in your area will help you to streamline your work more effectively. The more courses you complete in your area, the less daunting and time-consuming will be your preparation for exams in the specialized area. The more secondary literature you read prior to the dissertation stage, the more familiar you will be with the major critical discourses to which your own project will contribute. The more primary reading you do prior to teaching that first upper-level course in your area, the less time you will need to spend constructing a syllabus or preparing lesson plans. For these and other reasons, you should make it a point to take nearly every single course in your field.

2. Do take seminars in other disciplines. As an English literature scholar, I can say unequivocally that the most influential and informative course I took in graduate school happened to be in the Department of

History. Especially extra-departmental courses directly related to your field have the ability to influence not only how you perceive your primary materials but also the ways in which previous scholars have responded to them. Such courses, that is, encourage a relatively objective view of materials and practices you may no longer be questioning; they also offer you another discipline's terminology and research tools that can be applied in turn to the work you do in your own field. Note also that several graduate programs have begun to offer the equivalent of "minors". Consider adding a History or Women's Studies Minor or concentration to your Philosophy or literature Ph.D.

3. Do take seminars focused on periods or movements that immediately precede your own. Obviously it doesn't make much sense for an art historian to specialize in French neoclassicism without a solid knowledge of classical antiquity, Rococo, and perhaps the contemporary movement of Romanticism. Of course you should try very hard in the course-work stage of your graduate career to take as wide a range of courses as possible, to gain as comprehensive an understanding of an entire field as possible. It being *impossible*, however, to cover everything, focus first on those courses—within your department or in other departments—that will immediately enhance your understanding of your own field. Our neoclassicist art historian, for example, hopefully will take classes on many periods and movements not directly associated with neoclassicism; she will benefit immensely, though, from art history courses on Greek, Roman, and Hellenic art, as well as Rococo and Romanticism; from European history courses focused on the eighteenth and nineteenth centuries; from philosophy courses on the Enlightenment; and perhaps from theory courses on structuralism and post-structuralism.

4. Do take professional development courses and publication workshops. If you are reading this book, you're probably not the sort of person who would shy away from such useful opportunities, and this fact speaks to your wisdom. Surprisingly many students feel either that they do not need such "common-sense" courses or that they would be admitting their own deficiencies by taking them. I would simply stress that there is nothing commonsensical about writing a scholarly article or delivering a conference paper. In fact such skills are highly scientific, meaning that they follow very specific rules and patterns and improve as a result of practice and diligence rather than intuition or innate intelligence. Furthermore, such courses offer rare opportunities for discussing a range of topics that one simply cannot be expected to understand without years of accumulated experience in the field: what is the job market actually like? What is a teaching portfolio? How does

one write an appropriate cover letter to a journal editor? Learning the answers to these questions now will prevent you from wasting time later, and the confidence and savvy you will gain from this sort of practical knowledge will pay dividends in other realms as well.

5. Do pursue an "independent study" with an appropriate professor. The two best reasons for doing so would be to study important material unlikely to be covered in a graduate seminar, or to study material covered by seminars but unlikely to be offered while you are enrolled in course work. A secondary reason for pursuing an "independent study" would be to create an opportunity for working closely with someone you believe might make an excellent dissertation advisor. Independent studies can be very useful experiences for students and enriching ones for professors, but you should also know that they can be quite burdensome for the faculty. In many cases, an independent study will require an effort on the part of the faculty member the equivalent of teaching a seminar—only she will get no official credit or even much recognition for her effort. Especially for untenured professors, therefore, independent studies are potential traps that must be entered into with considerable caution. Simple awareness of this fact should help you to avoid making inappropriate requests of your professors; for example, it would be inappropriate to ask a professor to guide an "independent study" on a subject that she teaches every other year. In this case, you'd be far better off waiting to enroll in a seminar with her.

6. Do not take seminars for the sole purpose of studying for exams. My colleagues and I pull out our hair (I have none left, actually) whenever we think about how many students—especially less experienced ones—select courses merely to study for MA or, worse, Ph.D. comprehensive exams. Such a practice is not only offensive to the ideals of a graduate education, but it is shortsighted and potentially self-destructive. First, no graduate instructor who deserves to keep his or her job takes into serious consideration examination content while preparing a syllabus. In other words, the Victorian novel course you sign up for *might* include a George Eliot novel, which *might* happen to be covered by an examination, which *might* happen to ask a question about George Eliot. Or it might not. But is it really worth it to take a course simply because it might prevent you from having to read *The Mill on the Floss* on your own? A word of advice on this point: if you find yourself lacking the energy to read a George Eliot novel on your own, leave graduate school now (see more on exams in chapter 7). Second, as suggested by the aforementioned recommendations, you should select those courses that contribute most directly to your knowledge of a specific area and to the discipline as a whole. The logical correlation is that if

you are choosing courses to prepare for exams, you are likely doing yourself (and the college that might hire you) a disservice. In selecting courses, ask yourself two questions: first, which courses will enhance my knowledge of my field? Second, which courses cover material in which I am sincerely interested or about which I know too little?

7. Do not take courses that fail to require a longer research paper or project. While there are a few exceptions to this rule (introductory language courses, for instance, and MA courses with shorter weekly writing assignments), in general a course without a longer research paper (i.e., 15 pages or more) is not a useful class. One can always read material in one's own time, after all. The job of graduate instructors is not only to introduce students to representative material in a given field; it also is to help those students to find their own voice in the larger scholarly conversations about that field. Time spent writing a final five-page "response paper" or a short essay exam does not help one to find *anything* except the answer to why one has not yet published an article. Be sure to ask instructors about assignments before signing up for a class; if the final assignment happens to be a short essay exam and the class does not require a paper, ask a friend enrolled in the course for a copy of the syllabus and read the material over the summer. But do not waste your time taking the course.

8. Do not avoid professors or courses because you have heard that they are "hard" or "demanding." As a graduate student and a future professor, you should demand that your professors be demanding and hard. I found almost *without exception* in graduate school that the professors I learned most from happened to be those professors with the most frightening reputations. What this really meant was that they required more than the bare minimum of effort from their students, that they gave honest—not "inflated"—feedback on assignments, and that they behaved as though their classes were the most important ones in the world. If you're willing to confront your own anxieties, you will find that such passion tends to be infectious; and if you allow yourself to get swept up in it, you'll never settle again for mediocrity or complacency in the classroom. Further, the harder you work in your seminars, the more efficient your scholarly life will be later on.

9. Do not audit courses except under exceptional circumstances. To "audit" a course usually means to sit in and complete the readings, but not to complete any of the written assignments. I include this advice in this section because many graduate students select particular courses *only because* they have received professors' permission to audit them. Audits cannot stand in for credit courses, and they are worthless on a transcript, so usually students take them as a fourth or fifth course.

It stands to reason that if a course's material is important enough for a student to justify a considerable amount of additional reading—which will cut down on the time available for teaching and writing in other courses—then the course should be taken for credit. I can imagine only one scenario according to which an audit might be a good idea: in the case that a student *who has completed course-work* has an opportunity to take a class, not previously offered, which focuses on dissertation-related content. Otherwise audits should be avoided. Finally, should you ever choose to audit a course, regardless of the reason, avoid assuming that it's okay to perform less seriously than you would in a seminar for credit; that is, do not assume that it is ever acceptable to skip readings, miss classes, or remain silent during class discussions.

MEASURING STUDENT PERFORMANCE

As an undergraduate you likely got the gimmick early on: do the reading, show up, say something intelligent every few class periods, and write competently. *Magna cum laude* or *summa cum laude*? In graduate school, though, it's often difficult to know how you're doing or even if you're doing adequately. Partly because of the ridiculous grading system now in place in most graduate programs, according to which a "B" equals an "F," students have a hard time understanding what constitutes exceptional or mediocre or subpar work. The section that follows offers one faculty member's honest sense of the differences. I should state for the record that when it comes to calculating final grades for graduate students, my decision is based mainly on the student's final paper since it is the culmination of an entire semester's work, though other factors can drastically affect the calculation. But how faculty members assess your overall performance has far greater consequences than the grade you receive in their classes. These faculty members will be consulted when decisions are made about your qualifications for admission into the Ph.D. program and various internal grants; their conversations about you will impact how other current and future faculty members see you; and they may even be asked about you by colleagues at other universities when you are on the job market. Your goal, therefore, should be to erase all doubts in your professors' minds about the seriousness of your work ethic, the sincerity of your collegiality, and the quality of your work.

Participation

My basic belief is that a professor should have to restrain graduate students from speaking out passionately about the subjects under discussion. I've learned that, practically speaking, this is an unrealistic expectation, but I maintain the principle in every course I teach. Respecting shyness is, of course, both noble and decent, but unlike undergraduate instructors, graduate faculty members are responsible for encouraging all students to speak up. As a graduate student, you have chosen to enter an elite profession, the implication being that you might have something valuable to contribute to it. It is incumbent upon you, therefore, to show your actual ability to contribute something, however difficult it may be for you at the beginning of your career. Remember that no academic employer (and no future colleague) will wish to hire someone who will remain silent at department meetings, and they will extrapolate from your silence much about your teaching abilities and your intelligence—fair or not.

Most professors will conclude similarly that in-class silence signifies a lack of engagement, curiosity, or worse, failure to read the material. Make it a priority, therefore, to show your professors and classmates not only that you have read the material but that you are thinking actively about it. If you are engaged and still find intolerable the thought of joining a debate in class, consider the following strategy: based on your reading of the materials to be discussed in class, work out ahead of time two or three questions or observations. Write them down in your notebook, and be sure to share them with the class when appropriate. It goes without saying that if you are not engaged or curious about the material—and especially if you find yourself skipping readings—you probably should begin looking for another career.

Also make an effort to be considerate to your classmates. Try not to talk only to the professor, even if what you're saying happens to be in response to a specific question. Speak to everyone at the table, and work to reference and acknowledge what others have already said. Don't hesitate to ask your classmates to clarify a point and, by all means, don't shy away from challenging them to push their ideas further. A highly functioning graduate seminar should operate as an impassioned conversation where all of the regular rules of etiquette apply. Act intelligently, but show a willingness to learn; be challenging but polite, and people will respect you.

Work Ethic

Speaking up in class is not the only way to show how seriously you regard your work. Chapter 1 discusses the importance of approaching your graduate education as simply the first stage of your professional career; studying and learning is your job (not to mention a tremendous privilege), and like lawyers, doctors, and construction workers, you must work extraordinarily hard in order to develop and succeed. Estimates suggest that full-time professors work an average of about 60 hours per week, the workload being heavier prior to the earning of tenure. It seems logical to conclude that graduate students should be working at least as hard as their professors.

Office Hours

Unless every assignment she turns in happens to be perfect, any graduate student who fails to use office hours over the course of a semester surprises me. Because consultation and conversation are such an important part of the research process (see chapter 5), and because you need to demonstrate that you are an engaged, thinking individual, you should plan to meet with your professors several times each semester. My advice would be to go early to discuss your paper topic, however inchoate, to go mid-semester to update your professor on how the project is coming along, and to go again toward semester's end to discuss the developing structure of the paper. Such meetings send positive messages about your level of organization and degree of professionalism, and they help you to ensure that your work is in line with the professor's expectations.

Be careful not to overstay your welcome. The problem here is in knowing how to strike the right balance between dependence and independence. On the one hand, you must demonstrate your engagement in the class, your interest in the professor's feedback, your desire to know her better, and even your awareness of your own ignorance. On the other hand, your professors will expect you to be far more independent than their undergraduates, and they will have less patience for trivial matters or immaturity. Until you know a professor well, handle office hours the way you would handle an interview; know what you wish to discuss, be organized and absolutely professional about it, and, most important, be sensible about reading the signs that the meeting has come to an end.

Extracurricular Activities

Many professors arrange and encourage seminar participants to attend "optional" or "extra" events, such as film viewings, museum trips, guest visits and talks, even cocktail hours and luncheons. The basic assumption of such professors is that students who choose to pursue an academic life *should be* interested in intellectual activities related to the larger field they've chosen to study. Suffice it to say, you will disappoint your professors by failing to show up for academic events, and you may offend them by failing to show up at their end-of-the-semester pizza parties. Again, show yourself to be appreciative, collegial, and engaged.

Competence (i.e., the Bare Minimum)

Some, *usually* less experienced graduate students ruin their chances of being admitted into Ph.D. programs by acting immaturely or discourteously—usually because of ignorance about the ways graduate programs are run. In other scenarios, students' reputations are damaged among faculty members (always remember that your professors talk about you when you aren't around), which jeopardizes the students' chances of securing strong recommendations or, worse, of completing the degree. Consider the points that follow, then, as a few of the unwritten rules of graduate study:

Attendance: To paraphrase King Lear *very* loosely, "Never, never, never, never, never . . . miss a graduate seminar." I missed only one seminar in my graduate career. When I called the professor from a hospital bed, heavily doped up and in quite a bit of pain, he inquired suspiciously, "Which hospital" and demanded that I bring him a doctor's notice upon my release. An inconsiderate fascist, yes, but he sent a very clear message about the manner in which absences were perceived in my program. Obviously one can imagine a few legitimate excuses for missing a class—serious illness or bodily injury, the death of an immediate family member or close friend, and so on—but you should go out of your way to attend every class. Because a normal graduate seminar meets only about 14 times, one missed class is significant, and absences tend to be taken very personally by most graduate instructors. Should you need to miss class, try to notify your professor ahead of time, stress that you understand that absences are unacceptable, and make it a point to attend every other class that semester.

Incompletes: Incompletes are impractical for students and annoying for professors. While, again, there are exceptions to every rule, requesting an incomplete sends the message—to professors, Ph.D. admissions committees, and perhaps to search committees—that you have failed to manage your time very well. They usually force professors to return to grading papers during holiday or summer breaks, valuable periods of "free" time for working on research; it is safe to say that most faculty members resent anything that cuts into this time. Further, incompletes require professors to fill out rather tedious paperwork in order to process grade changes. For students, an incomplete can become a terrible psychological burden, potentially ruining holiday and summer breaks that would be more wisely used for working seminar papers into articles or conference presentations. In order to avoid creating more work for your professors and yourself, get your seminar papers done on time.

Assignments: Find out what the minimum requirements are for each assignment, and exceed them dramatically. "Just trying to pass"—an infuriating phenomenon that, nonetheless, is somewhat logical in today's quasi-vocational undergraduate education system—makes no sense whatsoever in humanities graduate programs. The point of the Ph.D. is to acquire the highest level of knowledge and expertise in relation to a particular subject matter; to consciously settle for less than your best possible work is to insult the degree and the very purposes of a liberal arts education. The most basic of the unwritten rules of graduate study in the humanities, therefore, is that you should never do simply what you are *supposed* to do; you should always do more.

Assessing Grades

Grade inflation is an old problem in universities. Rarely does the scientific indicator of an "average" performance—the "C"—actually represent the median grade for a particular class in the humanities. If the situation is troubling for undergraduate instructors, it is downright absurd for graduate faculty. In the institution in which I currently profess and the one in which I completed my Ph.D., grading breaks down in the following, undeniably strange way: an "A" means excellent; an "A−" means satisfactory or "B"; a "B+" means "C"; and a "B" means "F" (and, in the case that it is earned by an MA student, it also means "do not admit this student into our Ph.D. program"). While each discipline and each university has its own unique code language, I have yet to encounter any system more logical (or illogical) than the one I've just described. You will need to find out as soon as you arrive on campus what grades mean in your particular program.

A few more comments are warranted about this one, however. Master's students should keep in perspective the legitimacy and acceptability of an "A−." One of the most difficult aspects of grading graduate students is the oftentimes-serious disparity between the maturity and experience of Ph.D. students on the one hand, and MA students on the other. Generally speaking, master's students cannot be expected to perform on the same level as Ph.D.s, though some exceptional ones manage to do so. As an MA student, you should aim for an "A" in every course, but understand that an "A−" will not harm your chances of being admitted into the Ph.D. program, and consider a "B+" or "B" a warning sign that you must work much harder.

If your professor fails to offer written feedback about how your work might have been better—whether you have earned an "A" or a lower grade—you have the right to inquire. Always try to be accountable for your own shortcomings and respectful of the professor's point of view; do not complain or whine about your grade, but do express your concern and seek to learn how you might improve your seminar performance. Few professors mind offering students constructive feedback about how they might improve their writing, and most have fairly objective standards of what constitutes an "A" or a "B" paper. Students should understand that mediocre and even poor grades sometimes serve as helpful indicators of a problem that can be easily corrected. The quality of one's work can improve considerably as a result of an honest teacher's feedback.

Obviously Ph.D. students should be aiming for an "A" in every course, especially those focused in their own areas of specialization. While it is true that very few search committees will base a decision to hire you on your grades or transcripts (publications and evidence of strong teaching is much more important), "B"s stick out in a job file, and an "A−" will make it difficult for a professor to recommend you with a straight face as "the best of the best." Most important, though, a grade below an "A" sends the message that a Ph.D. student has yet to perfect one or another of the skills that will be vital to her success at the assistant professor level. Work to master the seminar system early in your graduate career to prevent the appearance of any red flags later on, when you can least afford them.

THE SEMINAR AND PROFESSIONAL DEVELOPMENT

As this book has stressed from the beginning, informed students understand the importance of time to completion, and they seek ways to tease out the connections between each seemingly separate phase of the Ph.D. process. At first glance, the basic value of course work for professional growth appears so obvious that it hardly seems worth discussing; seminars teach students to conduct research and construct article-length papers, which allows them, in turn, to write dissertations, present conference papers, and publish articles. This sounds great, even simple, but the level of abstraction and generality probably offers little to help students in actual practice. The next chapter focuses more specifically on the relationship between seminar papers and articles, and chapter 9 treats the relationship between oral reports and conference papers. In what follows here I briefly discuss the value of the seminar for dissertations and teaching.

There are considerable advantages to knowing one's dissertation topic prior to the final year of course-work (see also chapter 8). The final five or six seminars can then usefully double as opportunities for increasing one's expertise in the chosen subject matter and, even better, for drafting dissertation chapters. While it won't necessarily be possible or even desirable to use every single seminar in such a way, you should make it a point to look for connections between your topic and the material covered by each seminar. Let's pretend, for example, that you've decided to write a dissertation on "Labor Union Rhetoric in Twentieth-Century American Literature." You have six courses left before examinations. The two courses on twentieth-century literature offer obvious opportunities; although one focuses on British literature, it allows you to research the differences between American and British labor politics and rhetoric. While the paper you write for the British literature class is unlikely to form a chapter, the information you've turned up will figure prominently in the introduction of your dissertation. The nineteenth-century American literature course also allows you to work out introductory material—namely, to discover the roots of the rhetorical traditions upon which your dissertation will focus. You've also chosen to enroll in a history seminar on the industrial revolution, which offers you a valuable, interdisciplinary perspective on your subject matter. In the end, you decide simply to enjoy your "Shakespeare" and "George Eliot" seminars and to take a momentary rest from the dissertation topic.

This somewhat idealistic scenario nonetheless offers a sense of how you might lop months—even a year or longer—off the total amount of time you spend writing your dissertation. Such a well-planned course of action can realistically result in the following products: (1) a bibliography of primary and secondary works vital to the composition of your dissertation and ready for incorporation into your comprehensive examination reading lists; (2) one or more drafts of dissertation chapters; (3) a group of faculty advisors—whether official committee-members or not—who can offer a valuable first round of feedback on the project; (4) conference paper material that can allow you to "market" your project while it's still in progress; (5) and finally, the accumulation of enough introductory material to make writing a dissertation prospectus a timely and rather easy matter (which is what it should be [see pp. 48–49; 173–76]). By cutting down on the time required to complete the steps in between course-work and the dissertation—that is, the comprehensive examinations and the prospectus—and by drafting chapters before the ABD stage, our hypothetical student has streamlined her work most effectively and made the absolute most out of her seminars.

Graduate seminars can also help students to make more smoothly the difficult transition to teaching, especially upper-level courses. One of the basic issues here is organization. The materials you accumulate over the 3 or 4 years in which you are enrolled in seminars will serve a number of useful purposes down the road—both teaching and research related. Be sure to save them. As always, effective filing systems vary from individual to individual, but I would recommend that you take, at the very least, the following few steps: first, buy one or more large, three-ring binders, and store all paper materials inside—the syllabus, handouts, your own written work, and especially photocopies of secondary readings. Label each binder clearly and place them on your shelves or inside your filing cabinets. You'll be surprised by how often you'll return to these materials later on, especially in the semester (or on the night) before you have to teach a new subject (see figure 3.1 on page 68).

Also be sure to take careful notes on, or in, the books you read for each seminar. Since you're likely to teach from the same editions you used in graduate school, or at least to consult those editions, records of your own initial responses to these texts or of your classmates' and professor's in-class comments often prove extremely valuable later on. You might also make it a point to write down questions about the texts that seem appropriate for future classroom discussions. Personally speaking, I rarely ever kept an actual notebook in a graduate seminar, but the detailed marginal notes and queries in my books helped me to survive my first few attempts at teaching Spenser and Milton to eighteen-year olds. Similar to the ability to write on a particular topic, the ability to teach complex subjects depends upon the effective accumulation of knowledge and relevant experience; for the sake of your teaching and research, you should work to preserve, organize, and keep accessible the knowledge and experience acquired in every one of your graduate seminars.

Chapter 5

The Seminar Paper

Almost every serious graduate course in the humanities culminates in a final written assignment that I refer to in this chapter as the "seminar paper." Because the seminar paper both allows you to demonstrate your knowledge of the relevant course material and prepares you for the difficult tasks of dissertation writing and scholarly publishing, it might accurately be understood as the *sine qua non* of your academic training. Nonetheless, for most graduate students, confronting the seminar paper each semester is akin to launching an arctic expedition without a compass or a map; you may have some sense of where you want to go but painfully little guidance about how to get there. Although there exists no universally applicable set of instructions for writing a successful seminar paper, especially across disciplines, you can take certain steps that will help you to master the form—steps which, in ideal situations, might even lead to publication. This chapter focuses on the perils and pitfalls of seminar paper writing—and how to avoid them. Since wise students approach writing seminar papers just as they approach writing articles, this chapter serves as a supplement to chapter 10, which deals with the publication process. The major subjects include:

- The value of emulation
- The construction of a reading list
- The organization of materials
- The note-taking process
- The formulation of an argument
- The context of an argument
- The evidence of an argument
- The importance of your personal voice
- The process of revising for publication

The Value of Emulation

Especially because of the myth in academe that "originality" should be the goal of all scholarly research, the educational value of systematic imitation often gets overlooked. This is unfortunate since savvy teachers and students have long recognized that imitation is a starting point for learning in many pedagogical systems. As we will see shortly, the academic definition of "originality" needs to be understood within certain highly specific contexts, but first it will be important to discuss how you should envision the seminar paper, a skill that will require a certain degree of familiarity with the form.

Because scholars do not publish seminar papers, we must look to article-length essays as the most appropriate models for the approximately 20-page papers we are typically asked to write in our graduate courses. And why shouldn't this be the case? Many professors, after all, specify even in their syllabi that papers should be understood as practice runs for scholarly publishing. For example, each of my graduate students discovers on his first day in my class that a "20- to 25-page, potentially publishable final paper" will largely determine the final grade for the course. Indeed, such language was the rule rather than the exception in most of my courses as a graduate student, and it goes without saying that any student enrolled in a graduate program for longer than a semester is likely to have encountered it before. Of course, many professors—clinging to the outdated and somewhat irresponsible "apprenticeship model" of graduate education (see pp. 5–6)—deliberately shun such language in both their syllabi and their classrooms. More often than not, such professors are doing what they believe is best for their students; because they assume that an emphasis on the professional development of graduate students is only damaging and premature, they do what they can to protect them from preprofessional pressure. Although many students do, in fact, feel overwhelmed by the emphasis in today's graduate programs on publishing and conferencing, ignorance about how to publish, rather than recognition of the need to publish, is probably the cause of their anxiety. The facts here are simple: avoiding the realities of today's academic market, which demands publication, may make you feel less anxious in graduate school, but you will feel considerably more anxious later on if you are unable to land a job because you have not published. Even if you are lucky enough to secure a position, you may find yourself laboring frantically to do what should have become second nature in order to produce the publications necessary for tenure.

My advice, therefore, is that you embrace the seminar paper as a means of preparation for scholarly publishing. One positive and somewhat paradoxical side effect of such an approach is the diminishment of anxiety as a result of an enhanced sense of purpose and direction. As one of my best graduate students confesses, envisioning papers as articles transforms them from hoops through which one must jump into serious and potentially useful exercises: "I found that I only achieved a degree of success when I began thinking of my seminar papers as pre-publication attempts rather than as papers to get finished for a class. I guess the difference in my mind has to do more with preparation for the paper than anything else—i.e. going to original sources rather than casebooks, knowing the range of scholarship in the particular field in which I am working, translating languages, finding the best (or standard) editions for each text, etc." Unsurprisingly, the student not only has turned in consistently excellent seminar papers; he has also published several articles as a pre-dissertation student.

Often what distinguishes excellent from mediocre seminar papers is the mature student's knowledge of what published work actually looks like. While professors and advisors can help you notice certain typical characteristics of published writings, the time that you spend in the library reading and thinking about how to emulate these essays will prove far more valuable. My student knows that he should seek out the most highly respected editions of primary works because respectable journals simply do not publish authors who use modern-English translations of *Beowulf* to research the poem. From reading journal articles closely, you will learn much about how authors formulate provocative claims, how they build upon the research of other scholars, what sorts of materials and methods they use to persuade their audiences, and how much evidence they bring to bear on their own arguments.

By focusing actively on more than the rhetorical elements of such articles, you will also learn much about the mechanics and even the politics of scholarly work in your field. How many pages long is the average essay in the top three philosophy journals? How large is the average bibliography in an essay on the development of American labor unions in the nineteenth century? Which scholars' names repeatedly come up in discussions of Moliere's writings? Which style manual, *Chicago* or *MLA*, tends to be most respected by editors of Spanish literature journals? Your ability to answer such questions will help you to avoid seeming naïve or ignorant about the way things work in the culture of your particular field.

The next section focuses on what and how much to read for a seminar paper, but we should reiterate some of the basic claims we have made. One key to your success in negotiating the seminar paper assignment will be your ability to keep in perspective the larger reasons for writing such papers in the first place, the most important of which is to develop your knowledge of and ability to construct the fundamental unit of written scholarly work: the peer-reviewed article. By imagining what your work will look like in one of the top journals in your field (rather than in a pile on your professor's desk), you will gain an immediate sense of how to proceed in your research, based on your evaluation of the essays that have already been published therein. This goal-oriented sort of approach is likely to energize you by giving you a sense of purpose and drawing out the importance of the course-related work that you do as a graduate student. It may even result in some unexpected professional achievements. Systematic emulation of sound published work might with some degree of accuracy be called the starting point of all written research.

The editor of an excellent journal told me when I was an MA student that I should be spending at least two hours in the library each week simply reading recently published volumes of the top journals in my field. I would enthusiastically pass along his excellent advice to you.

THE RESEARCH STAGE

Once you have decided to pursue a particular interest, you are ready to begin your research. One of the factors that makes seminar papers more difficult to write than articles is the fact of uncontrollable deadlines. Although professors also have deadlines—for conference proposal submissions or article revisions, for example—they are more free to begin research projects when their interest in a subject happens to be piqued; graduate students are usually asked to select a topic on which to write on the first day of class (often before they know anything about a subject) or to turn in a proposal for a final paper midway through a semester. One positive aspect of such practices, however, is that graduate students are more often able to conduct research without bringing too many damaging, a priori assumptions to bear on their eventual conclusions. In ways, such research is purer than the type practiced by many scholars who look for ways to prove what they already believe to be true.

The research stage can be likened to Dante's journey through the heart of the inferno itself: only after descending directly into the depths of hell can the curious pilgrim eventually see the light of heaven visible on the other side. The analogy is useful only because it establishes the heroic and sometimes terrifying experience of confronting the well-known thinkers and ideas that have come before us. The specific sort of research we do depends largely on the fields we have chosen to study, but research across the humanities always begins with an engagement of the scholarly heritage of ideas pertinent to our subject matter. It follows that the older or more established the practices or texts we are studying are, the more time we will need to spend in this initial phase of our research. Organized graduate students, therefore, do not wait until the final weeks of class to write their final papers; in a sense, they begin on the first day.

Constructing a Bibliography

On the first day of class, you sign up to write on Milton's *Paradise Lost*. Now what?

Research in most disciplines today begins with one or another of the elaborate electronic databases that have only just begun to transform the academic research landscape. A simple search for *Paradise Lost* on the MLA Database—the best search engine for scholars of classics, English, comparative literature, linguistics, and the modern languages—turns up as many as 2, 355 citations. Other important databases include: the Getty Index for art history; L'Anée Philologique for classics; Historical Abstracts, ERIC, and America for history; Philosopher's Index for philosophy; and Women's Studies International for women's studies. Remember that such databases are not flawless; MLA, for instance, will turn up *most* of the available criticism on a particular text, but it does not record scholarship published before 1963, and often even recent publications fail to show up. Furthermore, searches are sensitive and, therefore, require a certain degree of ingenuity. You may find only 416 works on "Early Modern and Women," but another 500 or so turn up when you type in "Renaissance and women." The initial process of building a reading list, then, might begin with electronic databases, but it should always be supplemented by more traditional research techniques: searching your library's online catalogue, consulting notes and bibliographies in major works, discussing key texts with experts in the field (i.e., your professors). Finally, be sure to learn about more specialized electronic databases such as Early English Books Online (EEBO) which, in the case of *Paradise Lost*, would help you to locate relevant commentary published earlier than 1800.

Once you generate a preliminary list of works for consultation, you must decide what you can safely ignore because no one can possibly read all that has been written on *Paradise Lost* since 1667. In the case of less established or more focused research subjects—the court record of a particular witch trial in 1788 or the writings of a twenty-first-century philosopher—you may be able to cover *most* of what has been written. In any case, in deciding what to read, always cast your net widely (since articles on other subjects often contain ideas you can use), and always go directly to the source. A common mistake made by graduate students is to trust that what recent scholars say about previous scholarship is actually true or somehow indisputable. While Freud-bashing may be fashionable in modern scholarship, for example, it is incumbent upon you as a researcher to figure out for yourself whether or not Freud deserves to be bashed on a particular subject. The tendency to trust one's contemporaries, I should admit, is fairly understandable: for one, we are likely to accept ideas in circulation because doing so makes it easier for us to market own ideas; also, many common graduate-level assignments designed for wholly practical purposes, such as the annotated bibliography of the "last ten years of work," serve to reinforce the erroneous idea that modern scholarly practices have solved all of the problems that our primitive forebears were unable to overcome. We need to avoid conceiving the history of ideas in teleological terms, though, since such conceptions only lead to questionable scholarship. In the 1980s and 1990s, popular misinterpretations of Foucault's discussion of Panopticism, for instance, inspired numerous literary critics to explore similar modes of surveillance in relation to eighteenth- and early nineteenth-century texts despite the fact that Bentham's Panopticon was not actually put into practice until the late nineteenth century. Sean C. Grass has claimed that "[I]t is worth wondering whether recent scholarship focused upon surveillance has forged provocative links between the [Victorian] novel and the prison or only between the novel and Foucault."[1] If Grass's hunch is correct, we can say that such scholarly anachronism could have been avoided had more scholars familiarized themselves with Foucault's complex understanding of history or read more carefully about the history and fate of Bentham's original proposals for the Panopticon. The point is that you should always consider the *roots* of scholarly discourses and practices, not just the fruits that they continue to produce. If you are researching the relationship between madness and the nineteenth-century poetic imagination, you might begin by reading Plato's *Ion* and other classical texts. If you are researching twentieth-century feminist scholarship on *Paradise Lost*, you should look into what eighteenth- and nineteenth-century critics had to say about the poem's depiction of Eve. Only when you are confident that you know most of the important things that have been said about the subject will you be ready to begin writing.

There's a rub here, of course: how would you know that Plato's *Ion* happens to be a relevant text if you're not yet an expert on the subject? Perhaps the most difficult part of conducting research is figuring out which ancillary texts are important; a search on *Paradise Lost* will not necessarily turn up writings on Milton and women, which may be the subject you're interested in exploring. Here's where engagement of the scholarly heritage of ideas becomes crucial. Let's stay with our *Paradise Lost* example for a moment. We may not know much about seventeenth-century criticism of Milton's poem, but we do know of a famous essay by Gilbert and Gubar about Milton's apparent misogyny. The essay tells us that numerous, prominent women of the eighteenth and nineteenth centuries—including Emily Dickinson and George Eliot—recorded their negative feelings about Milton's Eve. Our research into Gilbert and Gubar's argument leads us to a counterargument by Joseph Wittreich that offers examples of other eighteenth- and nineteenth-century women who wrote more positively about Milton's Eve. Suddenly we have a list of about ten eighteenth- and nineteenth-century writings that searches on MLA and EEBO failed to turn up. Our bibliography is taking shape. As we begin to read both the primary and secondary materials, the bibliography will continue to grow while our interests continue to develop in more and more specific ways. The key now is figuring out how to manage all of this material and knowing when it's okay to stop reading (see also pp. 179–80).

Collecting and Organizing Materials

Since we have already discussed the importance of maintaining a thorough filing system (see pp. 66–67), we might be more specific here. Clear a shelf or two of your bookcases and take out two or three hanging folders. Label them specifically: since you are writing on Milton's Eve, you might consider beginning with three general categories: "*Paradise Lost* Criticism," "Milton and Women," and "Feminist Theory." Although the categories overlap somewhat, the maintenance of three folders rather than one will make it easier to locate things later on. Other researchers might prefer a different method such as organizing folders chronologically: a folder for writings between 1667 and 1800, one for writings between 1800 and 1900, and so on. Choose a method that works for you, and be willing to experiment.

At this stage you will need to recall books that are checked out and request other items through interlibrary loan. Both recalled books and interlibrary loans can take weeks to arrive so you should not wait to act. *Never* assume that it is acceptable to ignore works just because they are currently unavailable in your library. One of my colleagues at

Connecticut loves to talk about a former student who wrote a paper on the relationship between form and consciousness in the nineteenth-century novel. When he asked the student why the paper failed to include any reference to Ian Watt's seminal work on the subject, the student reported that the library's copy had been reported "missing." When my colleague asked why he had not bothered to order a copy from another university, the student became indignant, remarking that he "could not be expected to read everything." Because he had failed to consult the major work on the subject, though, he had no idea how major it happened to be (footnotes and bibliographies should have told him), nor did he know how much of his own argument had already been articulated by Watt. The "B–" he received on his paper could easily have been avoided had he been more thorough.

Once you have prepared appropriate space for storing materials, go to the library. The books you check out should be organized in some logical way on your bookshelves. I recommend to all of my students that they photocopy all journal articles and important chapters in books. (Be sure to include the title page and table of contents page so that you have hard evidence that the article involved is in this particular issue.) Many students learn the hard way that taking notes in the library may be cheaper than photocopying, but having to return to the library multiple times in order to quote material and then again to check those quotes tends to be more trouble than it's worth. By photocopying material and organizing it into the appropriate folders, you allow yourself easy access to it at all times, and you can mark up the materials as necessary. Especially if you decide to attempt revision for publication of a seminar paper, you can count on working with these materials for at least 2 years. Of course, if the articles are available online, you should simply download and print them out. In any case, avoid making unnecessary trips to the library, forgetting where you read something useful, or having to recall an item more than once in order to save a few bucks. Remember that in academe, as in other places, time is also money. By the time you are ready to begin reading, you should own a couple of very thick folders, and there should be some new books on your shelves.

Note-taking

While reading and note-taking strategies are as numerous and diverse as the people who practice them, you will maximize efficiency and increase retention by following a few basic steps. Because you may be confronting 50 or a 100 texts in a fairly short period of time, you can bet that you'll quickly forget or even fail to register much of what you are about to read. You can also bet that you will not be able to spend much time on every single text you have brought home. The following recommendations are offered to help you to wade through the most time-consuming part of the research process.

The Art of Skimming

Often my students confess feeling guilty about moving too quickly through books and even articles. You should recognize, however, that very few people ever read scholarly books from first word to last. Indeed, the most prolific researchers tend to be experts in the art of skimming. The introduction of a scholarly book typically is the most important part to read carefully. The introduction will tell you what the author is arguing and those texts he is covering. It will also tell you whether or not you should spend more of your time reading the book. Often subsequent chapters in scholarly books simply apply the argument offered in the introduction to various texts. To return to our *Paradise Lost* example, a book on the effects of Milton's republicanism on his poetry might be very useful for a paper on Milton's Eve. Read the introduction and the chapters on *Paradise Lost*. Consider skipping the chapter on *Paradise Regained*.

Index Searches

Every scholar appreciates a solid index. Often an index will tell you what you need to know even before you check a book out of the library or read its introductory chapter. Always spend some time leafing through a book's index before placing it back on your shelves. The index of the book on Milton's republicanism might remind you that Milton happened to write poetry about another fallen woman: in *Samson Agonistes*. Perhaps what the author has to say about Dalila will be useful to your consideration of what he has to say about Eve and Milton's basic attitude toward women. The index also might change your mind about reading the *Paradise Regained* chapter if it informs you that pages 300–310 treat Milton's depiction in that poem of Mary, the mother of Jesus.

Two Versions of Notation

Some people can't help themselves from writing all over the materials they read. Selective underlining and well-organized marginalia can undoubtedly prove useful in the research process, but I would stress the relative inferiority of in-text notes, which can be difficult to

navigate especially after some time has passed since an init
Furthermore, you cannot (or at least should not) write
books. I recommend using one of the two following strate

Buy some lined tablets. Beginning on a clean page, jot
basic publication information of the text you are about to read: include at the very least the author's name, the title, and the date. As you take notes, be sure to include the specific page numbers that will help you to locate the pertinent information. A brief sample from one of my own note pages will perhaps be useful (see figure 5.1).

Note that I have starred the site of the actual argument of the book, which may prove useful to me later on, long after I have forgotten all of the details. Also note that I've written down the title of a work I'll need to add to my bibliography, and I've made it stand out on the page. Develop your own system of shorthand to facilitate your research. After you have finished taking notes on a book, tear out of the notebook the relevant pages, staple them together, and insert them into the book before placing it back on your shelf. Before returning a book to the library, be sure to remove the notes and place them in the appropriate folder; consider photocopying and attaching the relevant material from the book first. If the notes refer to an article, staple or clip them to the first page of the photocopied document and place them back into the appropriate folder. You may wish to skip the tablets in favor of a single sturdy notebook, in which you will record all of your notes for a single paper. This alternative has the benefit of cutting down on loose sheets of paper that can easily get lost or disorganized. But it also requires that you keep the notebook with you at all times, which—if large enough—may limit when and where you can read.

Another strategy involves typing notes and other useful information directly into a word processing program. So that you don't have to sit in front of your computer while reading, consider making shorthand notations within each text and typing out more specific notes later on, after you've returned to your computer. Some scholars choose to record electronically the bibliographical information

---

L. Knoppers, *Historicizing Milton* (Georgia, 1994)
—Milton scholars slower than other Renaissance scholars in applying NH [New Historicist] methods (3–8)
—A few precursors to Knoppers's work include (5–9)
**Book's argument and scope (10–12).
Add —Zwicker book mentioned on page 8
Quote —underlined claim on page 12

Figure 5.1 Sample notebook page

pertinent to each text they read and the quotations they intend to include in the final paper, both of which can then be conveniently cut and pasted during the actual writing process. Another advantage of electronic notes is that they allow you to perform quick and highly specific searches for material you've previously recorded, whereas locating information in paper notes invariably requires a certain amount of shuffling and discombobulation. As always, find a method that works for you and stick to it.

FORMULATING AN ARGUMENT

Allow your argument to emerge from your reading. By attending carefully to the dominant conversations about a subject that previous scholars have conducted, you will find yourself forming your own opinions on the matter. Once you have identified what you wish to say, and determined that it has not already been said in the same way, you will be ready to write. Let us outline this process a bit more thoroughly.

The philosopher, Charles Peirce, defines a logical, argumentative process called "abduction," which

> makes its start from the facts, without, at the outset, having any particular theory in view, though it is motivated by the feeling that a theory is needed to explain the surprising facts. . . . In abduction the consideration of the facts suggests the hypothesis.[2]

Such a process differs from inductive reasoning, which would begin with a theory based on one's general observations, or even deductive reasoning, which always begins with a hypothesis. In the scheme of things, abduction might be described as an ideal method for humanities scholars, though we would be remiss to ignore its practical limitations: we all bring conscious and unconscious assumptions to our readings, and these corrupt our ability to be objective. Nonetheless, we should try at least to pursue the sort of pure research method Peirce is recommending. Whereas a deductive procedure might begin with the contaminating *assumption* that Milton is a misogynist, and an inductive conclusion leap to the transcendental *claim* that he is one, an abductive procedure would consider the facts: first, the primary writings themselves and second, what others have said about them. Only then would a relatively unbiased argument be possible.

Moving from this rather abstract consideration of the argumentative process, we might consider more carefully some practical recommendations for formulating a strong and useful argument. Notice that in all of the following types of scholarly argumentation, the researcher's goal is to solve a particular "problem."

1. The Controversy Paper: one of the most common forms of scholarly argument is the claim that purports to end a controversy or debate. Whereas a certain group of writers have argued that Milton is a misogynist, others have gone so far as to call him a proto-feminist. You have analyzed the relevant materials, and you have formed a view that can be backed up by evidence. You are ready to weigh in on the subject.

2. The Textual Crux Paper: for years readers have pondered the meaning of an ambiguous, unclear, or even a missing part of a given text, whether a poem, an oral expression, or a nineteenth-century police blotter. Or perhaps one recurring, but fairly cryptic word in a text catches your attention. Your research leads you to a strong conclusion about the meaning of the problematic text or term, and you set out to prove that your conclusion is valid.

3. The Gap in Scholarship Paper: in reading the scholarship about a particular subject, you are struck that no one has said anything about a related and seemingly important matter. You decide to widen the scope of the conversation. When I was struggling to find my own dissertation topic, I began by researching what made me most curious at the time: Renaissance conceptions of the human body. I was pleasantly surprised to learn that despite all the scholarly attention—even obsession—that the subject had generated, no one had ever analyzed Renaissance literary conceptions of health and exercise. As a result, I decided to write a dissertation about sport and exercise in Renaissance literature. The practice of locating "missing" conversations in scholarship can lead to significant research at the seminar paper level and, eventually, at the dissertation and publishing stages. You should constantly remind yourself, though, that the mere absence of discussion about a subject does not validate the importance of that subject. My decision to work on sport, for instance, would never have been acceptable had I argued simply that "Sport is an important subject to study because no one has yet studied it." We always need to consider the possibility that a particular subject has never been studied because it happens to be a boring and unfruitful subject. The key to making this type of paper work is your ability to say very specifically why a previously ignored subject should, in fact, be studied.

4. The Historical Contextualization: in recent years, the process of contextualizing practices and texts historically has been central to scholarship in the humanities. Perhaps your consideration of

contemporaneous documents or cultural practices helps in some way to clarify the meaning of a particular work or explain its provenance, immediate reception, or influence on other contemporary texts, people, and/or events. For example, reading seventeenth-century marriage manuals or even Milton's own writings on marriage might shed useful light on your inquiry into Eve's relationship with Adam.

5. The Pragmatic Proposal: in this sort of essay, you are more interested in praxis than theory for its own sake. Perhaps you have determined that Milton is neither a misogynist nor a proto-feminist; he is simply ambivalent about Eve, and you decide to write an article that demonstrates how highlighting such ambivalence in the undergraduate classroom can bring *Paradise Lost* to life for your students. It should be acknowledged that some professors will not accept this sort of essay in a graduate seminar, but you are likely to practice such a form of argumentation at different stages of your career. (This very book is a form of the pragmatic proposal.)

6. The Theoretical Application: Many graduate students seek to apply a theoretical approach—feminist, Marxist, or microhistorical, for example—to a text or other cultural artifact. Such assignments became popular in the 1980s and 1990s, decades that witnessed the ascendancy of high theory in the academy. While theory has always been and remains a crucial part of what scholars do, we should be careful about how we *apply* theory in our readings. Setting out to prove that Marxism always works is no less problematic than attempting to prove that Marx's writings shed no light on literature or culture at all. Such an approach entails that we present a highly subjective, ideologically constructed or historically contingent idea as an objective truth. Furthermore, your job as a scholar is to say something valuable about a complex subject, not to support or validate what someone else (Marx or Greenblatt or Foucault) has said on the subject. The latter makes you a disciple, not a thinker. However, if you determine that Marxist ideas and terminology help you to *articulate your own argument*, do not hesitate to appropriate them for your own purposes.

These six paper types by no means describe the only forms of argumentation that humanities scholars practice, but they offer a fairly comprehensive idea of the range of approaches you might take. Notice that in all of these cases, the construction of an "original" idea is not necessarily your goal. Rather than trying to invent the wheel, you might think about how to reinvent it; consider how your ideas contribute to a scholarly conversation, how they widen our understanding or expose the limitations of well-established ideas. As editor William Germano testifies, "The good news is that editors aren't really looking for what's radically original. Even the most experimental works of fiction are experimental within a recognizable context and history. What editors do look for is the new angle, the new combination, the fresh, the deeply felt or deeply thought."[3] With this in mind, ask yourself how you can solve a particular problem posed by a text or answer a question raised by previous scholarship or even your own reading.

After you have decided what to argue, ask yourself one more important question before proceeding: "so what?" Force yourself to explain why your argument is important or useful. Remember that the fact that "no one has ever looked at this before" may only mean that it is not worth looking at, not that it *should* be looked at. At this point, you should begin to share your idea with other people such as your classmates and especially your professor. This initial feedback is often as useful as the more detailed feedback you will receive later on. Seeking advice from others does not reveal your inadequacies or lack of independence; it suggests your maturity, your knowledge that research is never created in a vacuum, and your willingness to exhaust all available resources. Your colleagues will let you know how well you have articulated the problem you wish to solve and how persuasive you are in trying to market your idea. The next section of this chapter focuses in greater detail on how you might make your claims matter.

THE WRITING PROCESS

Once you have established a claim and decided that it is worth pursuing, you are ready to begin writing. Since many books describe methods of organization and outlining in the compositional process, I do not attend to outlining per se; instead, I focus on the basic rhetorical elements of a scholarly paper. I break these down into three activities: articulating your argument, situating your argument, and proving your argument. Such an approach is in no way intended to suggest that skipping an outline is a good idea. In fact, the three-part structure I offer here is partly designed to help you to construct logical outlines prior to actual composition.

Articulating Your Argument

Tell your audience what you wish to prove in your paper. Though this advice may sound so obvious as to be superfluous or condescending, I find myself reading paper after paper that lacks a clearly articulated claim. Sometimes the paper simply lacks an argument. At other times, the paper is so poorly written or jargon-laden that the argument cannot be identified. Regardless of the reasons for the problem, nothing is more annoying to readers, who should never have to ask themselves on page five or twenty why they are wasting time reading an unclear paper about Melville when they could be rereading *Moby Dick*. By establishing up front a contract with your reader—"you will read my paper in order to learn X"—you ensure that your reader's goals are commensurate with your own. Furthermore, everything that you say should be offered in the spirit of fulfilling this contract with the reader. By forcing yourself to say in absolutely clear terms what you are trying to do in any given piece, you will also help yourself to focus on the task at hand and avoid being sidetracked by digressions.

Do not wait to state your claim. Whether you are writing a ten- or a thirty-page essay, your audience will always appreciate knowing what you are up to within the first page or two. Consider your own reading practices. When you pick up a magazine or a newspaper, you look immediately to the table of contents or the titles of stories. Magazine and newspaper articles tend to have titles that explain what they are about, allowing you to make an informed decision about what to read and what to skip, but titles to academic essays can only hint at the complex arguments they contain. Your introduction, therefore, should grant your audience the same sort of decision-making ability conveyed in a magazine's table of contents. Since your professor has no choice but to read your essay, thinking of him as your primary audience may make you complacent in your writing. As the practice of trying to emulate a journal article would suggest, you would be better off imagining an audience of skeptical scholars who may or may not be interested in the specific subject about which you are writing.

The ideal length of an introduction will differ from writer to writer and paper to paper, but you should aim to convey your major claim within the first few pages. Think of your introduction as an abstract that forecasts the larger structure of the document. In figure 5.2, an entire introduction is offered in the opening paragraph.

Whereas the first few sentences work merely to establish the subject under discussion—*The Compleat Angler* as a sporting treatise—the middle section of the paragraph tries to establish the problem that the author will attempt to solve: that critics have failed to consider adequately what type of sporting treatise *The Angler* happens to be. This second step is crucial to the success of a scholarly argument because the announcement of the problem implicitly addresses the "so what" question. The eventual delivery of the actual claim, in the final sentence of the paragraph, suggests precisely how the article will try to

> Izaak Walton's *Compleat Angler* is certainly the most successful sporting treatise ever written. Never out of print since the first edition of 1653, the *Angler* ranks only behind the Bible and the Book of Common Prayer as the most frequently published work in the English language. Traditionally characterized as a simple pastoral dialogue by an equally simple, even accidental, author, the *Angler* has more recently been viewed as an allegorical protest against the precision of the Interregnum; as Steven N. Zwicker argues, the book "gave classic expression to the culture of sequestered Royalism." While Zwicker and other literary critics have helped to reveal the *Angler*'s general political context, however, no scholar has done justice to Walton's complex and highly specific engagement of official Interregnum policies regarding sports and pastimes. Most recent work has attempted to reconcile a traditional portrait of Walton as an innocuous, simple-minded countryman with a growing awareness of the political suggestiveness of his literary masterpiece. But Walton does more than passively evoke the mythological image of a pre-Interregnum golden age; in fact, he uses sport quite deliberately and systematically to critique contemporary laws proscribing communal recreations.[4]

Figure 5.2 Sample introduction from published work

remedy the problem. It does so in just enough detail to inform the reader and perhaps to provoke his curiosity, but not so much detail as to make reading the piece unnecessary. Although the author has chosen to limit the introductory information to one paragraph, other options might have been pursued. Perhaps a three-paragraph structure would have worked: an entire paragraph announcing the subject, a brief one on the relevant critical tradition, and a more detailed paragraph on the author's claim and specific plan for backing it up. In any case, the reader knows right away whether or not to continue reading.

Situating Your Argument

Since you need to make clear why your argument is significant, it is important to establish that you are contributing to an established, relevant scholarly conversation—not merely talking to yourself. By saying that you need to "situate your argument," I mean that you need to make clear how your argument fits into the larger history of ideas. At this stage, demonstrating your engagement and understanding of previous scholarship becomes crucial.

Depending on your discipline and the sort of argument you are writing, you will be able to organize the scholarship according to certain logical categories. Your paper on the *Paradise Lost* controversy regarding Eve, for example, might trace the tradition of reading Milton as a misogynist, then trace the tradition that reads him as a proto-feminist, culminating finally in an explicit statement of your view and how it contributes to the debate. A paper on a textual crux in the poem might begin by reminding readers of the problematic passage, then reporting how previous critics have dealt with it, then discussing the limitations of their interpretations, before concluding with your new reading of the passage. A historical consideration of the poem might show how previous scholarship has been contaminated by anachronism or ignorance about actual historical conditions before providing the new historical information that solves the problem. Regardless of which form of argumentation you choose, in other words, your contribution should be defined in relation to the previous scholarship on the subject. Consider the following few examples of "situating moves" from well-known, published pieces:

> There are currently two strains in criticism of *Paradise Lost*, one concerned with providing a complete reading of the poem . . . the other emphasizing a single aspect of it, or a single tradition in the light of which the whole can be better understood. Somewhat uneasily this book attempts to participate in both strains. My subject is Milton's reader and my thesis, simply, that the uniqueness of the poem's theme . . . results in the reader's being simultaneously a participant and a critic of his own performance.[5]

> [I]t will be essential to my argument to claim that the European canon as it exists is already such a canon, and most so when it is most heterosexual. In this sense, it would perhaps be easiest to describe this book (as will be done more explicitly in chapter 1) as a recasting of, and a refocusing on, René Girard's triangular schematization of the existing European canon in *Deceit, Desire, and the Novel*.[6]

> Determinists have often invoked the traditional prestige of science as objective knowledge, free from social and political taint. . . . Under their long hegemony, there has been a tendency to assume biological causation without question, and to accept social explanations only under the duress of a siege of irresistible evidence. . . . This book seeks to demonstrate both the scientific weaknesses and political contexts of determinist arguments. . . . I criticize the myth that science itself is an objective enterprise.[7]

In the first example, Stanley Fish seeks to advance criticism of *Paradise Lost* by attempting to reconcile two extreme camps of Milton critics. In the second, Eve Kosofsky Sedgwick explains how she will appropriate Girard's important study of triangulation for more specifically sexual–political purposes. In the final example, science historian Stephen Jay Gould interrogates and exposes the myth of objective science promoted by countless, previous scientists over a period of several centuries. In all three cases, then, the authors work first to identify common or popular ways of reading texts or cultural artifacts, and then they attempt to change how we read them.

A traditional way of describing the differences between professional scholarship and graduate work—especially between published books and dissertations—is to acknowledge that graduate writers tend to spend a good deal more time situating their arguments in relation to previous scholarship (see also pp. 182–84). There are several reasons why this is the case: first, unpublished or seldom published authors lack the sort of ethos that allows them simply to offer an argument without contextualizing it in a highly detailed fashion. For this reason, graduate instructors sometimes make them work to establish this ethos by demonstrating how much research they have done on a particular problem. Also, the relative lack of confidence characteristic of inexperienced scholars tends to result in overcompensation. The unfortunate result of this scholarly version of what Harold Bloom calls the "anxiety of influence" is the conscious or unconscious subordination of the author's voice to the voices of his predecessors. The increasingly professionalized nature of graduate studies and the need of assistant professors to turn their dissertations into books long before tenure review has closed the gap between dissertations and published books. Because of the realities of the academic marketplace, the gap between seminar papers and published articles has also begun to close. In seeking to situate your work, avoid subordinating your voice to the point where it becomes secondary or simply gets lost. Never forget that the purpose of discussing previous work is to highlight why you and *your work* are important and necessary. See also below on "Finding your Personal Voice" (pp. 108–13).

Proving Your Argument

Ironically and relatively speaking, this is probably the easiest part of writing a seminar paper, and it demands the least amount of space here. Because your introduction should establish exactly what you will be trying to prove and the method you will employ in order to prove it; and because your "situating" section will establish why your argument is important; you should know exactly what to do—and your audience should know exactly what to expect—once the argument has been established and contextualized. In an experimental psychology paper, this might be the point where you would offer a description of materials and methods. In a history paper, this is the point where you would introduce newly discovered artifacts or information. In the *Paradise Lost* paper we've been discussing, this is where you would offer your close reading of Milton's poems or your reading of Milton's Eve in the context of contemporary documents about women. In any case, your goal now is to show, not merely to tell, your audience that what you claimed in your introduction is in fact the case.

As in the other sections of your paper, keep in mind the value of organizing systematically the material you will use to accomplish your goals, and do not hesitate to tell your audience how you have organized it. There is much to recommend the following, introductory statement at the opening of the "proof" section of your paper: "After a detailed analysis of contemporary English attitudes about marriage and marital relations, I will show how Milton actually seeks to free Eve from the suffocating constraints of the traditional Renaissance marriage." You have now established your task, which is to offer a comprehensive report of contemporary attitudes, to offer a convincing reading of Milton's Eve, and to draw out the connections between the two.

Imagining your seminar paper as a three-step process of articulating, situating, and proving an argument will help you to organize a massive amount of information and to convey your ideas in a clear and systematic manner. Keep in mind that these are general rhetorical categories, not mandatory structural requirements. In some cases, you might wish to blur the lines between the three activities or to reverse the order of your procedure. Once you understand what a seminar paper looks like, you can manipulate the form in ways that suit your personal writing style and rhetorical preferences.

Finding Your Personal Voice

Of all the anxieties revealed by questions that humanities graduate students regularly ask me, the most common one has to do with the fear that their scholarly voice is somehow different from—meaning less authoritative than—the voice of publishing scholars in their respective disciplines. Students rarely, if ever, use the word "voice" to express this anxiety; further, their questions take multiple forms: how do I say something that hasn't already been said? How do seminar papers differ from articles? How do dissertations differ from books? But I believe that anxieties about the authoritativeness of one's written voice are, in a nutshell, what they're all really about. The fact that the creation of scholarship is so mystified a process only exacerbates feelings many of us experience in our first days of graduate study, which together expose something like an "impostor syndrome": am I really prepared enough, smart enough, diligent enough to be here with these well-prepared, intelligent, and hardworking classmates of mine? Did I somehow manage to sneak by the admissions committee?

Am I now going to be able to fool my graduate professors too? In this sense, we might understand the practical anxieties we feel about locating our voices as existential ones as well. And I want to add a third type of anxiety to the list, which is a professional anxiety that what we do in the humanities is neither respected nor understood as being important by most people outside of our relatively circumscribed worlds. To sum up, we feel like we don't know what we're doing, in part because no one tells us what we're doing, and then when and if we figure out how to do it on our own, everyone else tells us it's good for nothing.

On several occasions at UConn I've been fortunate enough to teach publishing seminars and workshops. I meant well in the first incarnation of these classes back in 2004, but in retrospect, I was pretty ignorant about what students actually needed to know and how to go about teaching it. In one particular class, I was yapping on and on about how the majority of one's time writing should be spent on introductions when one particular student interrupted to ask a series of pretty basic, yet brilliant, questions: "How *do* scholarly introductions really work? Is there *any* sort of science to them whatsoever? And if conclusions are supposed to reiterate the argument, how *are* they different from intros?" Though I was able to stumble through a response to Molly's question that was slightly less than embarrassing, I was also pretty open about the fact that I too found these questions difficult, and I promised to bring better answers to class the following week.

The next day I pulled down from my shelves several books with particularly memorable introductions and conclusions; there was no other rhyme or reason behind my selections, which I thought an ideal way to proceed, so I wound up with a pile of works that included Stephen Greenblatt's *Shakespearean Negotiations*, Stephen Jay Gould's *Mismeasure of Man*, William Empson's *Milton's God*, and Richard Rorty's *Contingency, Irony, and Solidarity*. Admittedly, all of these books are somewhat unusual in their boldness, but again, my main criterion at this point was "memorable." Let's have a look at what these introductions do:

1. From Greenblatt, *Shakespearean Negotiations*, 1988
   I began with a desire to speak with the dead. This desire is a familiar, if unvoiced, motive in literary studies, a motive unorganized, professionalized, buried beneath thick layers of bureaucratic decorum: literature professors are salaried, middle-class shamans.[8]

2. From Gould, *Mismeasure of Man*, 1981
This book seeks to demonstrate both the scientific weaknesses and political contexts of determinists' arguments. Even so, *I do not intend* to contrast evil determinists who stray from the path of scientific objectivity with enlightened antideterminists who approach data with an open mind and therefore see truth. *Rather I criticize* the myth that science itself is an objective enterprise.[9]

3. From Empson, *Milton's God*, 1961
I am anxious to make clear at the outset, because the revival of Christianity among literary critics has rather taken me by surprise, . . . *I think* the traditional God of Christianity very wicked, and have done since *I* was at school. Most Christians are so imprisoned by their own propaganda that they can scarcely imagine this reaction; though a missionary would have to agree that to worship a wicked God is morally bad for a man, so that he ought to be free to question whether his God is wicked. Such an approach does at least make Milton himself appear in a better light. He is struggling to make his God appear less wicked. . . . That this searching goes on in *Paradise Lost*, *I submit*, is the chief source of its fascination and poignancy.[10]

4. From Rorty, *Contingency, Irony, and Solidarity*, 1989
This book tries to show how things look if *we* drop the demand for a theory which unifies the public and the private, and are content to treat the demands of self-creation and of human solidarity as equally valid, yet forever incommensurable. It sketches a figure *whom I call* the "Liberal Ironist." *I borrow my definition* from . . .[11]

Next, I pulled from my shelf the first scholarly journal I could locate, which happened to be a 2004 issue of *Shakespeare Quarterly*. It featured only two full-length essays, but the introductions looked very familiar:

5. From Kathman, ". . . Freeman and Apprentices in the Elizabeth Theater," 2004
This essay is organized as follows. First, *I present* some background about the London livery companies and their apprenticeship system. *I then offer . . . . [and] I conclude* the paper with a look at the career of John Rhodes . . . . Collectively, this information provides a fairly clear picture of a system . . . .[12]

6. From Poole, "False Play," 2004
I wish to shed a little darkness on the chess game that occurs near the end of *The Tempest*. . . .[13]

In spite of the essays' different subjects, disciplinary audiences, an methodologies, a pattern nonetheless emerges: all of the introductory paragraphs (meaning not the first paragraph of each text but the ones in which the topic is explicitly introduced) feature 1) a first- person singular perspective, 2) a profound sense of a personal wrestling with the past, and 3) a more or less clear "procedural map" for the setting out and proving of the argument.

Next, I wanted to determine whether any similar patterns define scholarly conclusion-writing. So I chose to analyze the four final paragraphs of essays I was teaching that semester in Shakespeare and Milton courses:

7. From Norman Rabkin, "Responding to *Henry V*," 1967
   Henry V is most valuable for *us* not because it points to a crisis in Shakespeare's spiritual life, but because it shows *us* something about *ourselves*: the simultaneity of *our* deepest hopes and fears about a world of political action. . . . The inscrutability of *Henry V* is the inscrutability of history. And for a unique moment in Shakespeare's work, ambiguity is the heart of the matter, the single most important fact *we must confront in plucking out the mystery of the world we live in.*[14]

8. From Greenblatt, "Invisible Bullets," 1988
   [T]he histories [i.e., Shakespeare's history plays] consistently pull back from such pressure. Like Harriot in the New World, the Henry plays confirm the Machiavellian hypothesis that princely power originates in force and fraud even as they draw their audience toward an acceptance of that power. And *we are free* to locate and pay homage to the plays' doubts only because they no longer threaten *us*. There is subversion, no end of subversion, only not for *us*.[15]

9. From Richard Helgerson, "Staging Exclusion," 1994
   No other discursive community *I consider* in this study was as far removed from the councils of power as the theatrical. . . . Yet because of that distance, their representations of England are at once the most popular and, in the case of those produced by Shakespeare, the most exclusively monarchic that his generation has passed on to *us* . . . giving that genre a singularity of focus that contributed to . . . the emergence of the playwright—Shakespeare himself—as both gentleman and poet.[16]

10. From Ann Coiro, "Fable and Song," 1998
    Milton knew that his poems would not stay safely tucked in the sheltering and appreciative Bodleian. They are out under the sky with the throng, the vulgar mob of readers. . . . Infinitely repeatable, Milton's poetics provide a deeply ambiguous guiding hand. Whether *we* will choose liberty or license in *our* own parody of Milton is entirely up to *us.*[17]

Comparing these paragraphs to those in the books quoted above and the *Shakespeare Quarterly* essays yielded—with one slight exception—the same results, suggesting the following pattern: conclusions featured 1) a first-person *plural* perspective (a royal "We" of sorts or, as it's called in reference to mathematics conclusions, the author's "we"); 2) a widening of the topic from an issue within a field to the field itself (from *Henry V*, say, to Shakespearean drama or from Milton's poem *Samson Agonistes* to Milton's poetics); and 3) a profound sense of interest in the future of scholarship and/or our very lives.

Once such patterns become discernable, one might think it easy simply to begin adopting a more scholarly voice, but this would be only partly true. In order truly to understand how scholarly writing in the humanities often works, we also need to understand the epistemological foundations of those obvious characteristics defining our conventions. What the introductory paragraphs demonstrate is an authorial attempt to carve out necessary space for one's own ideas, something I refer to above as "the situating move" (see pp. 105–07). The author works to situate him or herself within a larger conversation or in relation to an established paradigm. In the conclusion, on the other hand, we see something like an authorial attempt to redefine that paradigm. The author fights in the beginning for a space to speak, struggles, that is, for a place where his or her voice can actually be heard—"I want to argue," "I contend," "I wish to argue," "I try to expand"—and the fight is very much against the *burden* of the past, against a history of influential ideas. In the final move, the author tries to assess what he or she has in fact offered to that conversation and then reflects on how the altered paradigm will impact the future of thought on the same subject. The relatively unimportant individual of the introduction, as marked by the "I", has by the conclusion become a member of an important community, marked by the "we". This is in no way a conscious process for the vast majority of scholars, however experienced they happen to be. But the commonness of the pattern (there are many exceptions, of course), and the fact that most of us never are taught it, suggests the largely psychological component underlying traditional scholarly writing in the humanities: anxiety about how we fit in leads to a self-assertion of presence and worth (of self, of ideas) which, in turn, leads to the enhancement of a conversation traversing space and time.

The majority of graduate students I've taught have struggled at the start of their careers, and sometimes at the end of them, to master these patterns precisely because their anxiety is understandably greater than that of most of their professors. Early on, the anxiety tends to be so great that one defers too much to the voices of the past, to the established authorities, and so there isn't enough insistence on the ability of the author to make a difference in the conversation. Before taking these patterns back to my seminar students in that first publishing seminar, I asked them to bring to our next meeting a copy of their introductory paragraphs from the papers they were work-shopping. Of the twelve introductions, only two featured an "I", a first-person singular perspective, and most of the other students agreed that these two were unusually strong writing samples. Of the twelve conclusions, only one featured a first person singular or plural perspective; the rest all merely attempted to summarize what had already been argued, with little consideration of the argument's impact on the future; in other words, the papers were tellingly still focused on, or burdened by, the past—revealing a lack of confidence in the contribution just proffered.

Since teaching yourself to be confident is nearly impossible, you must think in more practical and professional terms: learn the patterns, theorize their psychological and epistemological foundations, and then hope that eventual mastery of the patterns will foster confidence in your scholarly voice. Since there is overwhelming evidence that writing constructs or constitutes thought, rather than translating or conveying it, it can be argued that the repetitive process of "forging" one's scholarly voice is likely over time to result in a real voice that requires little formal awareness on the part of the writer. Yet while your goal should be to make yourself a confident scholar, I also think that what may improve the psychological aspect of your anxiety is the simple recognition that anxieties should not necessarily go away unless you're the sort of person that simply stops asking questions. My guess is that even the most experienced and well-practiced scholars remain just anxious enough throughout their careers to continue struggling productively to find their most effective written voice for each individual paper.

FINAL PREPARATIONS FOR SUBMISSION

Another advantage of beginning your paper early in the semester is that you'll avoid having to finish it at 5 AM on the day it is due. Allowing yourself several days for revisions will ensure that your work is as meticulous and, therefore, as persuasive as possible. Allow time for separate stylistic, mechanical, and rhetorical revisions. Be sure to check quotes for accuracy. Format your paper according to the appropriate style manual. Check the syllabus one more time to make sure you have followed the professor's instructions. Now rejoice that you have made it through hell.

Revising for Publication

A former professor of mine liked to muse that "no one ever publishes an essay that is filed away in his desk drawer." While you should never submit an essay for publication before you and your advisors believe it to be ready, you also should never underestimate your ability to transform a strong seminar paper into an article. I learned this lesson firsthand when, six years after writing a seminar paper, I decided to remove it from my desk drawer, dust it off, and revise it; I was pleased to learn several weeks later that it had been accepted by a top literary journal. The point is that while most of the seminar papers we write tend to wind up in our desk drawers or even our garbage cans, many are capable of a greater fate.

In reading the graduate papers that I receive each semester, I never cease to be amazed by the quality of the arguments that they contain. But as I explain to almost every graduate student I teach, publishing success depends as much on how an argument is presented as on the argument's quality, and presentation is more the result of hard work than imagination. The logical conclusion is that since most students are capable of formulating a solid argument, they should eventually be able to publish their seminar papers by continuing to work hard on them after classes have ended. Of course, not all seminar papers are worthy of publication; you should go out of your way to learn from your professor whether he honestly believes the paper to be publishable. You should also demand specific feedback from your professor about what sorts of revisions are likely to lead to publication. But you should take very seriously a professor's encouragement to continue working on a paper, and you should try not to wait six years to begin revising it. In academe, holidays and semester breaks are ideal times for performing such revisions.

After revising according to the suggestions of your professor, ask whether he would be willing to read the piece again. Be prepared to make further revisions based on a second round of feedback. If the professor has offered to read it again, do not hesitate to go another round. If not, seek out other experts in your department and ask whether they'd be willing to look at an article you're thinking about submitting to a journal. Only after you have exhausted all local resources and done everything possible to perfect the paper should you reformat and submit it to a specific journal (see pp. 220–25). Although you can't be sure that your work will be accepted for publication, you can be certain of at least one thing: it won't get published sitting in your desk drawer.

CHAPTER 6

TEACHING

Most of those first-year graduate students afforded the opportunity to teach worry far more about stepping into a classroom than they do about beginning their own course-work, which is perfectly reasonable. Course-work entails that you continue being a student, albeit a more focused and hardworking one; teaching requires that you redefine rather completely your position vis-à-vis the university and academe more generally. The difficulty of this transition is heightened by the fact that you are still a student yourself—still attending classes, still being evaluated by professors, still stressing out about your own in-class performance. On any given day, in a matter of several hours, sometimes minutes, you will move from a classroom in which your work has been analyzed, even interrogated, to one in which you are suddenly the analyzer of 20 or more students. How can you possibly be confident about your ability to teach others when you are reminded everyday how much you still have to learn?

This chapter aims not only to answer such practical questions, but to do so in a way that effectively situates teaching in the context of your larger academic career. Make no mistake, a strong teaching record is absolutely *crucial* to your success; quite simply, humanities Ph.D.s are not hired because of their research alone, regardless of what you may hear from feckless and bitter teachers in your graduate program. As we have seen, an extraordinarily small percentage of American colleges, about 10 percent, are classified by the Carnegie Foundation as Research Universities, which means that a majority of search committees will be *at least* as focused on your teaching record as your dissertation topic and research prospects. Even research powerhouses understand that in an academic job market where supply far outpaces demand, there simply is no reason to hire persons uncommitted to excellence in teaching.

What follows, then, is designed to help you make a solid case for yourself as a dedicated and effective teacher. Since there are many good books on the subjects of pedagogical theory and therapy, I focus largely on the more practical issues pertaining to teaching in colleges and universities, which you will need to think about if you are to succeed in the coming years. Some of the issues addressed in this chapter include:

- Contending with the stress caused by teaching
- Understanding the relationship between teaching and research
- Designing syllabi and lesson plans
- Preparing lectures
- Grading
- Building an impressive teaching portfolio

By learning how your role as a teacher figures in the larger context of an academic career in the humanities, you can alleviate the anxiety associated with teaching and begin to experience the exhilarating confidence, the intellectual stimulation, and the tremendous personal rewards that can result from teaching in the higher education system.

A note on terminology: in order to avoid the inaccurate and politically offensive term, "teaching assistant," I use the terms "graduate teacher" and "graduate instructor" in all discussions of graduate students who teach their own classes. I use "assistant" only when graduate students are paid to *assist* professors or adjuncts.

REALIZING YOUR QUALIFICATIONS

Few situations are more terrifying, even for confident and experienced individuals, than facing a group of strangers who harbor two particular assumptions about you: first, that you know more than them; and second, that you are in a position to judge them because of what they do not know. Whereas the first assumption highlights a qualitative difference between the minds of the teacher and the students, the second defines a power relationship that is the inevitable result of this difference. Although both assumptions have firm bases in reality, they reveal one of the most significant challenges of teaching, which is the need for the instructor to break down (or prevent from developing) the adversarial relationship between teacher and student. If students feel threatened by or anxious about the relationship, they will not allow themselves to learn. The problem often is exacerbated when graduate teachers show up on the first day of classes. Not only does the teacher look young and inexperienced and therefore *more* threatening, but he or she may also try to compensate for inexperience by asserting authority in ways that only worsen the situation.

The only point I want you to understand here is this: you *are* qualified to teach a section of underclassmen. The students in front of you have little or no college education. You have a BA, maybe an MA, and, presumably, a very solid academic record, or you wouldn't have been granted entrance into graduate school in the first place. Probably you have gone through an orientation program that has begun to prepare you for what you are about to face. Further, you have countless resources all around you in case you should need advice or reassurance. Think about what you knew and thought before college, and think about what you know and think now, after 4 years of reading, writing, and learning to think critically. Remind yourself of the difference every time you feel nervous about stepping into that room.

The Importance of Teaching

I try to remind myself every time I walk into a classroom that good teachers changed my life. I do so because I teach much more effectively when I'm focused on the educational mission of the university. It's not always easy to stay focused, of course. Two major obstacles to effective teaching happen to take the forms of anxious undergraduates and your own psyche; perhaps the most insidious form, though, happens also to be the most seldom discussed one: colleagues or, worse, mentors who bash their students and teaching in general. That curious breed of primate we discussed in chapter 1, the "hall-talker," loves especially to stand around in the hallway—not getting anything done—talking about how stupid and hopeless her students happen to be. Usually fresh from a session with Allan Bloom, she latches onto anyone's ear she can find and proceeds to explain all that is wrong with this particular generation of students (technically speaking, usually the same generation of which she is a part). Or perhaps you've run into "research boy," you know the guy who is kind enough to remind you every time you see him what a waste of time teaching happens to be, usually right before he asks you to join the revolution. If it weren't for him, you wouldn't even know that most schools "only care about research" or that "teaching doesn't really matter."

Though I joke, the problem happens to be quite serious. Teaching requires a significant emotional investment, a daily surrender of the self that inevitably leads back to disappointment and self-doubt. The life of the teacher, like the life of the researcher, is defined by extraordinary highs and dreadful lows that people in most occupations would find surprising or even intolerable. Dedicated teachers deal with the lows because of the thrill of the highs, and the latter would be probably impossible without the former. Those lows constitute particularly dangerous moments, though, precisely because they have the ability to render all too appealing the obviously inaccurate and unproductive generalities of hall-talker and research boy. They comfort us by offering convenient explanations—scapegoats, really—for our own shortcomings, the failings of the larger educational system, or simply of circumstances that happen to be beyond our control.

The most basic thing I have learned in almost 15 years of teaching in the American higher education system is that our students are many things, but they are not stupid and they certainly are not helpless. If you adopt the attitude now that they are, you will be bitter and angry for the rest of your career since, as a humanities Ph.D., teaching will always be an important part of what you do. Rather than dismissing students as fools, you need to try to understand why certain ones act as they do, whether lazy or indifferent or even defiant; realize that it is your job as an educator to address the problems your students face—not simply to identify them.

Notice that I am not denying how frustrating some students can be. There will be at least one student in just about every class you teach who is unreachable or, rather, who won't allow herself to be reached. She will sit in the back of the room—when she bothers to come to class—and she will roll her eyes, or sigh out loud, or simply stare out the window all hour long. And though you may be getting through to every one of the other 20 or 50 students in the class, this one problematic individual will haunt you when you lie down to sleep at night. The first thing you will do is question yourself: "what am I doing wrong?" The next thing is that, in order to protect yourself, you will begin to rationalize the student's behavior. The student *must* be an idiot or a jerk, after all. Soon you will begin to hate the student and, if you obsess just a bit more maniacally, you will project this student's behavior onto everyone in the class and begin to hate all of the students. Soon you will be standing in the hallway with your new friends lambasting your stupid and helpless students.

Educational theorist Mike Rose characterizes student defiance and indifference as parts of a "powerful and effective defense" mechanism that helps neutralize the frustration caused by recognition of one's own ignorance or feelings of powerlessness:

> Reject the confusion and frustration by openly defining yourself as the common Joe. Champion the average. Rely on your own good sense. Fuck this bullshit . . . books, essays, tests, academic scrambling, complexity, scientific reasoning, philosophical inquiry.[1]

Ironically, what Rose describes as a coping mechanism for countless students applies with amazing accuracy to those graduate students, adjuncts, and professors who love to bash college teaching: reject the frustration by defining yourself as indifferent to your students. Champion mediocrity for everyone. Forget challenging essays, tests, and so on. Retreat into your own mind. And most important, ignore the fact that you've become like the person in the back of the room who merely whines and rolls her eyes.

In my mind, the most wonderful part of a career in academe is the balance one is encouraged to strike between near solipsism and utter selflessness. In what other world can someone embrace simultaneously something so idiosyncratic as a specialized research career in the humanities and something so socially vital as a teaching career in higher education? We should never forget that the privilege to pursue our own research interests grows out of and, in many ways, depends on the importance of our social importance as educators. By the time you retire 30 or 40 years from now, you will have impacted in one way or another literally *thousands* of young students—their ways of thinking about and understanding the world in which we live. As an absurdly well-educated individual in whom a great deal of faith has been entrusted, your responsibility is to struggle everyday against the cynicism that would prevent you from effectively and enthusiastically imparting what you know.

For those who succeed in educating their students, there is tremendous joy. Of all the jobs for which I am responsible as a university professor, nothing gives me more pleasure than teaching. Nothing else can compete with the "high" I experience every time I teach a good class or the pride I take in knowing that I've contributed to a student's intellectual growth. Most people know on some level or other how important teaching is. Few realize how wonderful it can be for those who take it seriously.

PEDAGOGICAL THEORY AND PRACTICE

The cliché goes something like this: whereas high school teachers possess all the skill but none of the content, college teachers possess all of the content and none of the skill. There is a certain degree of truth underlying many clichés, and this particular one is no exception. Beyond orientation and, in some cases, a first-semester teaching practicum, future academics receive surprisingly little pedagogical training. The negative effects of this obvious problem are felt by the students whom we teach and, if they go unaddressed, they also will plague us in countless ways throughout our careers. For example, you can be sure that the vast majority of questions you will face in job interviews will pertain to teaching undergraduates. Many otherwise strong candidates blow their chances of getting a job by being unable to answer some version of the following, common question: "Which writers on education have most directly influenced your teaching?" Sometimes the questions are more specific: "Where do you fall in the Hirsch/Freire debate?" "Is Allan Bloom simply a crank, or is he onto something?" "What do you see as the limitations of bell hooks' conception of class in the classroom?"

I want to make absolutely clear that while your department probably *should be* responsible for teaching you how to answer such questions, it probably won't do so. This means that you'll need to do a lot of reading on your own. You might also consider taking a few of the following steps in order to begin building an adequate understanding of pedagogical theory:

Read the *Chronicle of Higher Education:* Yes, I've already given you this advice in an earlier chapter. The point is worth reiterating here because the *Chronicle* features several teaching-related articles per issue. Whether the article happens to be on the developing centrality of technology in the college classroom, one obscure teacher's views regarding critical pedagogy, or a famous pedagogue's recent work, you will benefit immeasurably from exposure to the terminology and methodological focus of cutting-edge pedagogical writings. For the same reason, you might consider subscribing to a journal that's focused on teaching in your particular discipline.

Consult Your Professors: When appropriate, simply ask your professors about those educators and educational theories that have most influenced them. Not only are these discussions likely to be provocative and stimulating, but they will also help you to expand your reading list. By consulting numerous people in your field, you'll start to pick up on who the big-name theorists are in the discipline—who's in, who's out. Such "inside information" will be invaluable on the market.

Seek Out Educational Presentations and In-house Colloquia: In most universities you will have numerous opportunities to attend talks on a variety of teaching-related subjects. Seek out appropriate lectures in the department of education, and make it a point to attend colloquia by professors in your own department.

The most important reason to take these steps has to do with your own edification, of course. An immersion in pedagogical theory will help you to discover and learn to articulate those methods that work best for you. There's no reason why skill and content can't be happily married.

MANAGING A COLLEGE CLASSROOM

There is no comprehensive user's manual for new teachers. And since new instructor orientations tend to focus mainly on pedagogical and policy-related matters, new teachers typically are asked to enter the classroom without much practical knowledge of what they are likely to face there. While there is no adequate substitute for experience and trial-and-error learning, the following section is designed to offer some advice about matters too rarely discussed in official forums.

Making a Syllabus

At no point in my graduate career did anyone ever speak to me about the process of constructing a syllabus. Because we are exposed to syllabi as undergraduates, and often because we are given one to use before our first semester of graduate school, people assume that experience alone has been an adequate teacher. Having come to understand the rhetorical and institutional importance of the syllabus, I consider such negligence to be quite extraordinary; simply put, the syllabus is one of the most important documents with which you will work on a regular basis. In this brief subsection, I break down the *necessary* components of a typical syllabus. You can also consult the syllabi in the appendix (pp. 291–95).

Personal Information

Name, office location, office hours, phone number, and e-mail address. If you prefer to be called Ms. or Mr. Johnson rather than Sue or Mike, make this fact known on the syllabus. If you decide to provide your home telephone number (I recommend that you do not), be sure to emphasize the latest time in the evening when students can call you without invoking your wrath. Next to your e-mail address, you may wish to provide a note on what constitutes appropriate e-mail inquiries. Do not allow students to e-mail you for information available on the syllabus; you have better things to do with your time than answer e-mails about when office hours take place.

Course Description

Think of the "Course Description" as the first impression you will make on your students. They will infer from your syllabus your level of enthusiasm, the seriousness of your expectations, and the style of your presentation. If you write a boring course description, they will probably assume that you are boring. On the other hand, if the description is merely sexy, with no real substance, they will think you are trendy or a pushover. The key to a successful description is the balance between being provocative and being serious. Figure 6.1 offers an example of a course description that reveals a lively, intellectually curious, and fairly demanding personality.

Materials

A list of all texts for which students are responsible, including handouts, films, and so on. If you require that students study particular editions of any texts, be sure to specify an appropriate amount of publication information.

Requirements

List all requirements, not just assignments. In other words, if you have an attendance policy, now is the time to mention it. Also make clear how you will weigh each assignment, as shown in figure 6.2.

The requirements section of the syllabus is a good place to send a clear message about the seriousness of your expectations, especially since most students will look to this section before any other. The best case scenario is that students unwilling to meet your expectations will simply drop the class.

In this course, we will study the rich literature of the English Renaissance in its historical, religious, and philosophical contexts. We will also explore the concept of "renaissance" itself: its usefulness, accuracy, and appropriateness as a descriptive term. The first reason to take the class is the fascinating literature we will be reading. From the epic power of *The Faerie Queene* to the sublime nihilism of *King Lear* to the witty satire of Donne, this literature will stimulate, shock, and change you. All semester long, we will be engaging with poetry, drama, and prose works that have helped to shape—for better or worse—our current ways of thinking and feeling. Second, English 221W will help you to see how the foundational themes and characteristics of English literature have changed over the centuries. Understanding how a particular literary motif is developed, revised, and even turned upside-down by different writers in different historical contexts will not only enhance the reading you do in other college classes; it will enhance the reading you do long after you graduate. Finally, English 221W will help you to cultivate strong rhetorical skills, both in your writing and in conversation. Through a number of short written assignments, you will learn to read a text closely, to formulate a powerful critical claim, and to back that claim up with supporting evidence. These skills will be vital to your success both in college and as an employee, in whatever context, after college.

Figure 6.1 Sample course description

Requirements	
Class grade (preparation, participation, attendance)*	15%
8–10 Reading quizzes	25%
Take-home midterm paper	20%
Film review	20%
Final exam	20%

* The professor does not take kindly to excessive **absences**. Because your understanding and engagement of the material depends on your presence in our class, I will take absences seriously. More than two, for any reason, is inexcusable. **Preparation** includes completing homework and in-class assignments, and having read the texts to be discussed in class. Please make it a point to turn off your **cell phone** before class begins.

Figure 6.2 Sample course requirements

Policy on Plagiarism/Cheating

In some recent surveys, as many as 50 percent of students have admitted to plagiarizing at least once in their college careers.[2] A much higher percentage of students claim that they would plagiarize were it not for fear of being caught. One would have to be seriously naïve to believe that plagiarism doesn't occur in every single writing class we teach. The internet has, of course, worsened a very old problem. My advice is that you send the *strongest* possible message in your syllabus that academic dishonesty will not be tolerated and that the maximum penalty for violators will be enforced. If you determine later that the crime was accidental or somehow less than egregious, you can lighten the penalty. But in order to protect yourself and your right to punish cheaters appropriately, you must establish that students were aware of the consequences of cheating. Find out the policies at your university regarding academic dishonesty, and work into the syllabus its official language, as shown in figure 6.3.

The final sentence is particularly important; some students are simply ignorant about the differences between plagiarism and quotation, and you must be willing to educate potential cheaters as well as punish actual ones.

Course Schedule

A good course schedule will not only demonstrate your organizational and planning skills; it will also communicate to students the internal logic of the course—revealing a narrative or at least a logical movement from beginning to end. Specify exactly what students are responsible for at every single class meeting. If the class is to read *The Return of Martin Guerre* over three class periods, it will be helpful for students to

Plagiarism: It goes without saying that you are responsible for citing any words or ideas that you borrow. Using material from the so-called Internet Paper Warehouses constitutes a form of plagiarism as serious as using someone else's paper (and is easy to discover). Plagiarism demonstrates contempt for your instructor, peers, and the purposes of liberal education. **If you are caught plagiarizing, you will automatically fail the course for violation of the student code and be referred to the dean of students for judicial affairs.** If you are uncertain as to what constitutes plagiarism, please consult the English Department's policies guide or see me outside class.

Figure 6.3 "Sample plagiarism policy"

know that you will cover material through page 75 on the first day of discussion. Also mark due dates for assignments on the course schedule.

Additional Information

Many instructors wisely include additional information on their syllabi according to the idea that all expectations should be clear. While you'll want to avoid making the syllabus too busy, you might consider including paragraphs on individual assignments, office hours, your attendance policy, and so on. Consider your syllabus a contract with students that details the obligations of both parties. And don't back down from enforcing those policies.

Lesson Planning

If you figure into the equation reading the materials you assign, we can say that lesson planning is the most time-consuming part of teaching—more so than time spent in class, grading, or in office hours. Your supervisors have no doubt told you that should spend no more than 20 hours per week on your teaching. Recognize that the reasons for this recommendation mainly are legal; operating according to the logic that office employees work on average only 40 hours a week (which no professionals do any more, by the way), administrators can tout the fact that so-called teaching assistants, who only teach 20 hours per week, are mainly students, not employees. As we discussed earlier, the classification of graduate students as employees would be detrimental to the university, which depends on graduate students' nonemployee status to prevent the formation of graduate student unions.

You will almost certainly spend more than 20 hours per week on each class you teach, especially early in your career. In order to protect time for research and service obligations, then, you must be as efficient as possible in your preparations for class. The key to such efficiency is long-term planning. In the profession, "new preps"—meaning the process of preparing to teach material for the first time—are widely and accurately regarded as extremely time-consuming affairs. Keep in mind while in the depths of despair that material is always easier to teach the second time around, but only *if you have already created detailed lesson plans.* A solid lesson plan can be used for years, requiring only minor tweaking to improve and update it. A poor lesson plan will have to be rewritten and may even have to be trashed. The lesson here is simple: work hard in the beginning and your workload will be greatly reduced later on.

So what defines a good lesson plan? Of course, different teachers will approach planning a lesson in different ways, but several considerations will ensure both that your classes are well organized and that your lessons will continue to be useful in subsequent semesters. Figure 6.4 presents an abbreviated lesson plan for a class on *Othello* that highlights the document's most important features.

To begin, the teacher chooses to include "housekeeping" issues so they won't be forgotten. Next, the teacher summarizes the major issues covered in the previous class, which helps the students to remember where they've been and sets a context for the new material. Once the teacher turns to the new material, she takes several important steps: first, she includes specific citation information for the passages she wishes to cover; as a drama teacher she is wise to include act, scene, and line numbers since it's likely she will teach from different editions of the play over the course of her career. Also, she uses boldface to highlight questions for students; the boldface helps her to remember to what end she's decided to analyze particular passages; in later courses, such reminders will be valuable. After the question, she provides several possible answers; now if students fail to respond or simply respond differently than she has anticipated, the discussion can be shifted in a preferred direction. Finally, the lesson includes information that sets up students for the following class.

The point is not to create a perfect document from which you plan to teach for the rest of your career. You will be as bored as your students by old, unrevised lesson plans. But revising and updating is preferable to writing from scratch. Rest assured that nothing is wrong with this particular practice. Personally speaking, I don't believe one can really nail a class the first time around; usually, it takes two or three trials to get it right. As long as you're willing to rework the plan once you sense it has become ineffective, you can feel confident teaching the same basic material for years. By noting on the lesson plans after each class what worked and what didn't, and filing them away in a

Lesson Plan for April 13, 2004

1. Mention film screening tomorrow night: 7 pm 163 CLAS. Write on board: "motiveless malignity"
2. Summary of previous class: *Flash up map again.*
   - A) We discussed the moral crisis the play forces its readership to undergo by making it complicit in Iago's treachery. Similar to the case of *Richard III* but probably worse in that Richard is cartoonish and somewhat absolute.
   - B) We discussed the correlation between geographical locations and psychological states, and especially the dichotomy between barbarism and civilization.
   - C) On a related note, we discussed the central issue of race in the play, both in terms of how the Europeans view Othello's black skin and his own self-consciousness about being black in a white culture. We left off by suggesting that Othello is **the tragedy of a man trying desperately not to become what white society tells him he should be, which is the Barbarian: 2.3.168–73. The fight against the Turks symbolizes an internal battle against barbarism**.
3. Continuing Discussion of *Othello*:
   - A) Iago is able to succeed by exploiting Othello's anxieties. **How exactly does Iago operate? What are his methods of destruction?**
     - 1) For one, he uses other characters' strengths or virtues to destroy them. There's a relentless irony to this method.
       - a) He recognizes that Othello's honest and open nature is a bit of a weakness: **2.1.288–89. Can appeal to it:** CF. **3.3.365–66, 370–78.** Here Othello's essentially virtuous personality is tantamount to naivete.
       - b) Does the same to Desdemona: **2.3.358–62. Makes strategy explicit**.
     - 2) Second, he taps into Othello's worst anxieties about his race: **3.3.257–66**.
     - 3) Implants concrete images of the deed in Othello's head: serves as the "Ocular Proof" that Othello demands: **4.1.1–5**. . . .

....

4. For next class:
   - A) Finish reading play.
     - 1) Think about Iago's final words. **In what ways do they conclude or continue Shakespeare's exploration of evil in *Othello*?**

Figure 6.4 Sample lesson plan

place where they will be easily accessible the following semester, you'll know exactly where to begin when it comes time to revise.

Fraternization

One advantage held by young teachers happens also to be a disadvantage: the greater ability of twenty-somethings and some thirty-somethings to relate to their undergraduate students—especially in reference to popular culture-related issues—may also lead some of those undergraduates to view the instructor as a potential buddy or friend. The problem is again exacerbated by universities' refusals to recognize graduate teachers as employees since the typical legal classification of TAs suggests they have more in common with college freshmen than they do with their professors. Your responsibility is to be aware of this danger and to emphasize at all times your professional role as an educator. And *under no circumstances whatsoever* is it ever acceptable to date or sleep with one of your students.

GRADING AND GRADE-INFLATION

A colleague who has read this chapter in draft form tells me that it is "politically incorrect" to confess how much I loathe grading. Since I've been unable to find anyone who actually likes grading (or simply doesn't dread it), I've decided to do the politically incorrect thing and ignore her feedback. For humanities professors who rarely can justify scantron-type evaluations, the process of grading papers, exams, and quizzes tends to be grueling, time consuming, and tedious. For graduate students in the humanities—who usually are asked to teach the least appealing and most rigorous courses in our departments—the grading process can be much worse. And to top it all off, in today's consumer-driven, corporate universities, teachers aren't even allowed the freedom to assess student performance accurately; if one wishes to avoid a dozen complaints by students and countless, time-consuming questions by the department (and sometimes college) administrators, one better learn to give students the grades for which they are paying.

The pressure on teachers to inflate grades, I would argue, is somewhat intolerable—in some cases diminishing significantly the joy of teaching certain classes. While observers such as Alfie Kohn have demonstrated that, statistically speaking, the problem is not that much worse today than 50 years ago,[3] anyone who has taught in a university setting for more than 10 years will tell you differently. When I began teaching college students 10 years ago, my supervisor informed me that an 80 percent, or a letter grade somewhere between C+ and B−, should be the average class grade. While this didn't make any sense, mathematically speaking, I thought little of it at the time. Today there are indications that the average grade in the humanities is closer to an 85 percent, or a letter grade of B. A *Boston Globe* study from 2001 shows that 91 percent of students at Harvard, America's most prestigious university, are graduating with honors.[4] My guess is that American students have not gotten smarter since 1995. Rather the grades "D" and "F" have mysteriously begun to disappear from most grading books as teachers have come to realize the serious disadvantages of being an honest evaluator of student performance. Worse, students seem more comfortable complaining about grades they don't like.

A demanding teacher will not only find herself bombarded by angry students socialized into believing that grades are like trinkets in a street fair—items to be haggled for—but she may also suffer worse consequences. Since students "shop around" for courses during the first two weeks of classes, upon showing students the syllabus she is likely to find her classroom empty or scarcely populated, at best. Further, she may find herself publicly slandered on websites such as "myprofessorsucks.com," where anonymous students are permitted, even encouraged, to bash their professors; almost invariably, nasty reviews on such sites are responses to tough grading practices. (Do yourself a favor and don't look at your online reviews.)

So what should we do? Throw our hands up in despair and give every student an "A"? Not an option for any serious educator, of course. Doing so would not only cause you to despise yourself, but also to be despised by your colleagues and even the students you are flattering. The key is to figure out how to maintain integrity as a grader, to figure out the methods that will allow you to grade fairly and accurately without making your life more stressful and difficult.

Practice What You Preach

As mentioned above, your syllabus can be a very effective ally in difficult classroom situations. A good syllabus will convey clearly and authoritatively your expectations and level of seriousness. The policies outlined in a syllabus are only as strong as the person upholding them, however. If you show yourself to be indecisive or unwilling to enforce the penalties for violations of your policies, students will perceive you to be weak, and they will be more likely to challenge your assessments of their performances. My personal approach is to establish on the first day of class that I have very high expectations for student behavior and work. The slackers and the frauds tend to drop immediately, leaving me mainly with those students who opt not to run from challenges. In short, you must work to establish right away the sort of professional relationship you find most functional in order to prevent problems later on.

Use E-mail Sparingly

While e-mail has made educators' lives simpler in ways, it also has led to the emergence of various problems related specifically to the grading process. The ease with which students can log onto their computers allows them to act impulsively—to communicate things over the internet that they would never say to someone's face. As soon as students learn they have earned a midterm grade of "C," for example, they can fire off an angry e-mail explaining that they have never "gotten anything less than a B before" (never believe this claim) and demanding to know why you are so mean. If you should stoop to the student's level by responding angrily, the situation can quickly get out of control. Don't allow this to happen. Explain in your syllabus that e-mail is not a substitute for office hours and that students with questions about graded papers or exams must make it a point to attend your office hours. Never suggest anywhere on your syllabus or in class that grades are negotiable. Keep the stress on the future, not the past; that is, encourage students to see you so that you can discuss how they might use their graded work in order to improve by the time of the next assignment. If a student should send you an inappropriate e-mail, you can then point to the policy: "Should you wish to improve your writing, I'll be happy to explain to you in office hours why you've earned a C. But I would also reiterate the point I make in the syllabus: grades are not negotiable." Now if the student decides to attend your office hours—if the student is *truly* interested in improving her work, that is—you both will be on the same page regarding what you can accomplish during the meeting.

Encourage Office Hours

One point of cutting down on e-mail is to encourage students to use your office hours. You will find that students are more likely to be respectful and focused in a one-on-one session than they are on the telephone or over e-mail. When discussing grades, your office represents an official or formal space that reminds students of the need to be professional. Such professionalism makes it easier, in turn, for you to respect the student and to stay focused on the task at hand. Since many students express hesitation about attending office hours, either because they feel intimidated by the teacher or skeptical about their teachers' sincere willingness to help them (too often for good reason), I use my syllabus to alleviate some of their anxiety:

> I urge each of you to take advantage of the opportunity to introduce yourself to me, to ask any questions you may have, to discuss future or current assignments, or to seek private instruction on specific problems with which you might be wrestling. I like students, I love teaching, and I promise that I don't bite. There is a too often unrecognized but undeniable correlation between students who tend to use office hours and students who tend to be successful in college.

As mentioned above, you should always keep your discussions with students focused on the future, on what they will need to do to improve. If a student should ever act inappropriately or angrily, which is rare, maintain your composure as best as you can, and ask (i.e., demand) that the student leave immediately: "I'm sorry, but I don't talk to angry students. If you decide that you would like to discuss what you need to do to improve, feel free to come back when you're a bit more calm."

An extreme situation, obviously, and one you will only rarely encounter. Usually meetings with students during office hours are productive, even pleasant experiences and one of the few opportunities college instructors have for actually getting to know the individuals whom they teach. By encouraging students to consult you prior to turning in an assignment, you'll send a clear message about your willingness to help them, and you'll cut down on the likelihood of a misunderstanding about grades down the road. Do not agree to read drafts unless you are given considerable notice and an appropriate amount of time to go through them. But do listen to students' ideas, however inchoate, and help them to develop those ideas into persuasive arguments even before they begin formal work on a particular assignment. If you are to convince students that each assignment is merely a single step in a much larger *process* of learning, you'll need to demonstrate your recognition of and dedication to that process.

Give Clear Assignments

The clearer the students' understanding of your expectations, the fewer problems you are likely to encounter regarding the grading process. Be sure to tell students exactly what you expect from them in the way of mechanics, if mechanics are important to you. For example, it makes less sense to say "The paper must be 4-pages long" than it does to say "the paper must be 4-pages long (double-spaced, times new roman 12 pt. font)." You may also wish to tell them that a standard margin is 1.25 inches. If you want them to format their papers according to the guidelines set out in *The Chicago Manual of Style*, tell them to consult that style manual.

More substantive issues are trickier, of course, precisely because a student's ability to judge what constitutes a "strong" or a "persuasive" paper/exam will depend on the student's maturity and experience. Teachers must be careful not to be overly prescriptive since the result is likely to be formulaic rather than original and creative work. Your responsibility at the assignment stage is simply to offer the most accurate terminology for describing your expectations. For example, if you ask students to turn in a three-page paper on Gayle Rubin's "The Traffic in Women," it will be helpful for them to know whether they should summarize Rubin's argument, analyze rhetorically Rubin's argument, or respond viscerally to Rubin's argument. (There are other options, of course.) Always read the assignment aloud in class, allow the students to ask questions about it, and invite them to discuss the matter further in office hours.

Provide Clear Feedback on Graded Work

You should always remember that you are an educator, not a judge. Especially on written work, grades without comments are useless to your students and likely to provoke frustration and even anger. In your written comments as elsewhere, your goal should be to focus students on what they will need to learn in order to improve for the next assignment. Written comments on graded work should not be thought of as justifications or even explanations of the letter or numeric grade. Rather they should help students to see what they have done well, what they have done poorly or not at all, and what they might do if they could do it again.

Like many teachers, I break my feedback down into marginalia and a summary-oriented statement at the end of the paper or exam. Throughout the document, work to direct the writer's attention to strengths/weaknesses by constructing a conversation of sorts in the margins. Whereas a cold declarative formulation—for example, "You need to revise!"—is likely to put students on the defensive and decrease the likelihood that they will learn anything, a more subtle approach may cause them to think more carefully about your suggestion: "Okay, fair point, but would it be more persuasive to consider X?" Whenever possible, try to cast critical feedback in the form of a question, which will be less intimidating and more thought provoking. In your summary statement, address the student by name and, as always, focus on what is necessary for improvement on the next assignment. Even if your marginal comments point to six or seven problems in the student's writing or logical thought process, try to focus your summary statement on the one or two most significant problems. Students should be able to take away from your comments both a sense of *what* they should try to improve and two or three concrete suggestions about *how* to improve. By teaching students how to think and write more effectively—rather than judging how well they already think and write—you'll communicate more persuasively the point that you are interested in their success and well being. The result will be a relationship defined by cooperation rather than antagonism, mutual respect rather than anger and distrust.

Miscellaneous

Just a few more tips about grading. Try to return graded exams or papers within one week of the time they are handed in. Students will appreciate your efficiency and organization, and you will relieve yourself of a major burden. While planning course schedules that will appear on their syllabi, experienced teachers learn to schedule due dates strategically, at times when their own work is less likely to be particularly intense. Finally, get into the habit of grading student work late in the day or at night. While your students' exams and papers will require your undivided attention, grading is less of an imaginative or critical-thinking oriented activity than a high-energy task. Your own research and writing activities demand that you be at your sharpest; my advice is that you prioritize accordingly.

While grading is not the reason most of us choose to become academics, it is one of the basic foundations of learning. By systematizing your approach to grading in ways that challenge the current consumer model of education—by approaching grades, that is, as signs of what students will need to do to become better thinkers and communicators—you can eliminate many of the unpleasant aspects of the process and transform grading into a rewarding and constructive activity.

EVALUATIONS

I have no doubt whatsoever that the state of higher education in America would improve overnight if colleges and universities were simply to end their reliance on computerized student evaluations. Not only would administrators be forced to discover an evaluative system that actually works, but they would also send the message clearly that students do not occupy the same position as customers at Wal-Mart. Computerized student evaluations are problematic for a few reasons: (1) because high evaluations tend to correlate with high grades. Evaluations are one of the primary causes of grade inflation, especially for graduate students who wish to be hired someday and assistant professors who wish to be tenured; (2) because evaluations cause teachers to avoid controversial subject matter in an effort to be liked by every student in a class; (3) because teachers will find it nearly impossible to translate the numbers into feedback they can actually use to improve their teaching; (4) and because evaluations mislead administrators into believing that more useful and accurate forms of evaluation are unnecessary.

Defenders of the status quo will call this exaggeration. They will say that the computerized evaluations empower students and root out ineffective teachers. In fact, they do neither very well. Although student evaluations of faculty teaching are taken very seriously at many small colleges and universities, most Ph.D.-granting institutions (i.e., the places you are likely to obtain your degree) are large research universities where teaching is viewed by administrators as less important than research. In addition graduate students are temporary workers or short-term investments. They must be evaluated for competence, not long-term excellence, and so their evaluations are likely only to be glanced at, if anything, by administrators and faculty supervisors. Undergraduate students at such research universities tend to be rather cynical about the purposes of the evaluations, having learned long ago that they do very little to ensure solid teaching, especially in business and science departments where evaluations are taken even less seriously than they are in the humanities. This is the bleak reality, which every young teacher should confront head on. By doing so, one might understand earlier that whatever benefits might come from computerized evaluations must be actively sought out by individual instructors willing to read between the lines.

Written student evaluations can actually be quite helpful, especially at the beginning of one's career, and they should be taken much more seriously. Not only can written comments point out weaknesses in one's teaching that can easily be improved and strengths than can be more fully developed, but the written comments will also form an important part of your teaching portfolio (see pp. 137–46). Because the standard questions on such forms are often overly vague ("How does the instructor compare to others at X university?"), make it a point to give your students prompts that require of them greater focus and specificity. Ask for their feedback on materials, assignments, even pedagogical techniques, especially those about which you are most uncertain. Be sure to ask them how they believe the course can be improved. Be careful to use positive formulations instead of negative or neutral ones ("In what ways did you benefit from the peer reviews," rather than "Were the peer reviews helpful?") And, most important, explain to students before you leave the room that you *will* read their comments and look to incorporate their feedback into other courses you will teach; explain to them, in other words, that they need not be cynical about whether their voices will be heard. By announcing your own willingness to learn and improve, you will teach your students a valuable lesson about humility, which should inspire them to offer more thoughtful and detailed feedback.

Written faculty evaluations, which usually take the form of "Letters of Observation," should also be read very seriously. Many universities require that all graduate instructors be observed at least once a year by a faculty member in the instructor's department, a practice that also applies to untenured professors. Unlike students, faculty members will have almost no larger context for understanding the material they observe you teaching, their observations being limited to a single visit to your class in the middle of the semester; as experienced teachers, though, they are in a position to notice immediately the strengths and weaknesses of other teachers, and they are usually more able than undergraduate students to articulate their understanding of a classroom dynamic. From faculty observers you should gain two important things: first, honest feedback about your teaching, which can be incorporated immediately; second, a written letter that can later be developed into a formal letter of recommendation for your teaching portfolio and your job application packet. Once you learn which faculty member will observe you, introduce yourself and explain that you look forward to receiving the professor's feedback. Try to schedule the visit for a day when you are teaching material with which you feel comfortable or which you believe highlights your strengths in the classroom. Provide the observer with a syllabus and any other materials that will help her to contextualize the particular class to be observed. Explain to students ahead of time that the class will be visited by an observer.

Do not change what you do on the day of the observation. Routine will keep you and your students on track, and besides, who wants feedback on something one does not regularly do? A certain amount of nervousness is normal, even healthy, but try to keep things in perspective. Every faculty member has at one time or another been in your position, and their feelings of empathy will cause them to look for the most positive aspects of your teaching and to give you the benefit of the doubt when things go less than perfectly. The wisest step you can take to assure a solid class is to pretend you are alone with your students. Once the class is over, explain to the observer what will happen in the next class or two, again in an effort to contextualize the observed class, and encourage them to give you honest feedback. Show a willingness to learn from more experienced teachers.

Prepare yourself psychologically for interpreting all three forms of evaluation—the computerized student evaluations, written student evaluations, and written faculty evaluations. Remember that all three forms are of *limited* value as indicators of how well you teach; anger, misunderstanding, or envy can taint severely the outcome of any one evaluator's response to your teaching. Negative comments and sometimes nastiness will stick with you much longer than positive feedback. But if you seek to draw on what's constructive and positive, to take seriously what's legitimately critical, and to ignore what's merely personal and downright unfair, you'll be able to glean much that is useful from the total evaluative process.

Prepare yourself practically for interpreting all three forms as well. The feedback you collect over the years will help you to articulate more objectively your own strengths and weaknesses as a teacher, which will be extremely valuable in the classroom and in interviews for jobs, grants, and teaching awards. As you will see shortly, the written data you accumulate will constitute an important part of your teaching portfolio, which will be crucial to you on the job market. The evaluative methods currently in place at most universities are by no means flawless, but they can serve useful purposes if understood and approached in the right way.

ASSISTING OTHER TEACHERS, OR THE *REAL* TA

Most professors give specific instructions to TAs, whose roles tend to vary dramatically from class to class, so I limit this section to three simple points of advice: first, ask up front what the professor's expectations are for your role in the class, including information about the number of hours you are supposed to work each week. If your TA contract demands that you put in 20 hours per week, you are personally responsible for tracking those hours. Keep clear records of all the work you do for the class. Do not complain if you happen to work 22 hours one week; the likelihood is that you will work 18 hours the following week. But if you feel that you are being significantly overworked or exploited, inform the professor of the situation; if the professor does not respond appropriately, consult the director of graduate studies. Second, be sure to explain to the professor how you wish to be addressed by the students. If "Miss Johnson" makes you comfortable, it won't do for the professor to call you "Sue." In other words, you may need to remind your professor from time to time that you are concerned about your ethos and would appreciate her reinforcement of your professional role. Finally, if the relationship between you and your professor is comfortable, request that you be permitted to teach a class or two or, at least, to deliver a presentation. Too often TAs are forced to do grunt work, which benefits the professor, of course, but not the graduate student. If the professor grants your request, ask her to be present during your presentation(s) so that she is in a position to write a letter of observation for your teaching file.

TEACHING PORTFOLIO

By the time you begin interviewing for professorial jobs, you will need to compile, organize, and, in some cases, generate a number of documents that, together, offer a detailed portrait of you as a teacher.

For the sake of emphasis, I will be blunt here: a solid teaching portfolio is yet another useful tool not only for distinguishing yourself from the crowd but for competing effectively against job candidates from elite universities. The Ivies, for instance, offer their students *relatively* few opportunities for diversifying their teaching, which is a significant disadvantage on the job market. Unwritten policy in one mid-Atlantic state university's English Department, for example, dictates that no candidate will be hired who has published fewer than two articles or taught fewer than ten course sections of more than one class. Considering the number of humanities candidates per available job in the United States it makes little sense for colleges and universities to risk hiring inexperienced or uncommitted teachers (*or* unproven researchers, for that matter). By constructing a concise but comprehensive portfolio, one that proves your experience and your commitment, you'll let search committees know you're anything but a risk.

Although the version of the portfolio that you will eventually turn over to job search committees will be approximately ten-pages long, your informal teaching file may be hundreds of pages long by the time you defend your dissertation. Deciding what should go into a portfolio, therefore, is a complex and sometimes painful process. In what follows, I break the portfolio down into five necessary parts: a teaching CV, the teaching philosophy, student evaluations, syllabi, and faculty/students observations of your teaching (i.e. letters of recommendation). As always, you should strive to be creative and add to this basic portfolio whatever other documents you think might enhance your portrait.

The Teaching File

As soon as you begin your graduate teaching career, you should make room in your cabinet for a large "teaching file." Into this file, over the years, should be inserted: a copy of every course syllabus from which you have taught; sample lesson plans from each course; sample quizzes, examinations, and writing assignments; photocopied papers demonstrating the sort of feedback you typically offer students; returned written and computerized student evaluations; documents demonstrating positive feedback from students (e-mails, letters, postcards, whatever); teaching-related letters of recommendation by faculty members, graduate student colleagues, and/or undergraduate students; and ideas/proposals for future courses or assignments.

You should continue to keep such a file even after you begin your first job as a professor and well into your professorial career. The materials it contains should always be accessible since you will need to rework your portfolio every time you go up for a promotion or are nominated for a teaching award. The file may also prove crucial should you ever need to defend your teaching record before a tenure committee, a university appeals committee, or even in a court of law. Think of the file as proof of your serious commitment to teaching and a diary of sorts.

The Teaching CV

By no means a standard form, what I call the "teaching CV" is a one-page document I worked up for my own job search that allows reviewers to gather in a glance the relevant *facts* about a candidate's teaching experience. See figure 6.5.

Begin by listing the courses you have taught, making sure that you provide enough descriptive information for readers outside your university. For instance, simply listing "English 15" will get you nowhere since people outside of your department will not know what English 15 is; opt for "English 15: 'Freshman Composition.' " Consider listing the number of times you have taught each course and the number of students enrolled, especially if you have taught large lecture courses as well as the typical, smaller courses. Be sure to make absolutely clear whether you taught the course on your own or assisted a professor. The best way to do this is to break up the teaching experience section into two categories: "Courses Taught" and "Courses Assisted."

Georgio Valenti

Teaching Curriculum Vitae

911 Oakwood Avenue
State College, USA 16806
Phone: 834/234-1647
gmv149@statecollege.edu

Department of English
116 Burrow Building
State College, USA 16806
Fax: 834/863-7245

Academic Employment

Teaching Assistant, The State University, 1997–2001
English 497: Literature and Film (1 section)
English 444: Chaucer (1 section)
English 129: Introduction to Chaucer (2 sections)
English 221: British Literature to 1798 (2 sections)
English 202B: Writing for the Humanities (4 sections)
English 202D: Business Writing (4 sections)
English 15: Rhetoric and Composition (6 sections)
***Teaching Assistant Award for Outstanding Teaching**, April, 2000. One of five recipients of this annual, university-wide award.

Teaching Papers and Publications

Published, "Reimagining Shakespeare through Film," in *Reimagining Shakespeare for Children*, ed. Naomi Miller. New York: Routledge, Forthcoming 20 pp. ms.
Published, "Teaching Meta-Theatricality through Film," *Shakespeare and the Classroom*. Forthcoming 2 pp. ms.
Will Chair, "Teaching Shakespeare on Film." Central New York Conference on Language and Literature. SUNY College at Cortland, NY. October 2000.
Presented Paper, "Scar Wars: Richard III and Darth Vader: Teaching Meta-Theatricality through Film." Workshop at the "Shakespeare Association of America Annual Conference," Montréal, Canada. April 2000.
Presented Paper, "The Concept of Meta-Pedagogy." Penn State Composition Program Evaluation Series. March 2000.
Guest Speaker. "Students and Teachers: Preparing for a Graduate Career in English," Juniata College, Huntingdon, PA. November 1999.
Discussion Leader: "Cultural Difference in the Classroom: Is Analytical Writing Gender-Biased?" State College Brown Bag Lunch Series Presentation. September 1999.
Workshop Leader: "Writing the 'Personal Statement' for Graduate School in the Humanities." English Graduate Organization. September 1999.

Figure 6.5 Sample teaching CV

List any teaching-related awards or honors. These may vary from departmental or university teaching awards to unusual privileges extended to you by your department (for example, permission to teach an upper-level class when such classes are typically off-limits to graduate students). Next, record any teaching-related service that you have performed. Have you ever helped out at the new teachers orientation?

Have you ever participated in a colloquium on a subject related to teaching? Put it down. Finally, list any publications or conference presentations on pedagogical issues. Such achievements demonstrate a real commitment to teaching and a willingness to bridge the gap between teaching and research.

The Teaching Philosophy

People too often tend to ignore the "philosophy" part, which is problematic. What the one-page teaching philosophy *should not be* is a description of the classes and books you have taught in the past. Rather, you need to think carefully about the pedagogical principles and ideals that inform and tie together every class you teach, however different those courses may be from one another. If your philosophy of education happens to derive from a particular educational theorist (e.g., Freire or Giroux), then share this information, since it will help your readers to ground abstract ideas in concrete forms. If your teaching philosophy derives from more personal or less direct sources, then consider coining a term or a phrase for it, which will serve the same concretizing function. In the example below (see figure 6.6), the author uses the somewhat unwieldy but descriptive term "meta-pedagogy" to characterize a personal approach to teaching.

Once you establish the principle or the theory, focus on praxis. Since often there is a significant gap in education between the pedagogical ideal and the practical reality, it's important to show how your ideas apply in actual classroom situations. Provide at least one example of an assignment or in-class practice that shows how your philosophy can be put to use. Avoid vague gestures in favor of clear, detailed statements; for example, rather than reporting that "I ask students to read many controversial books in order to foster debate," try the following: "by pairing the writings of thinkers as opposed as E. D. Hirsch and Paolo Freire, I encourage students to approach difficult educational issues dialectically." The more specific you are, the greater the likelihood that your readers will be interested and the more talking points you will generate for the interview.

Try to avoid clichés and buzz words that seem flat, stale, or unprofitable. "Utilize" is not a term that you should *use* since you're almost certainly *using* it incorrectly. I am quite certain that you do not have a diverse "plethora" of students. While it's probably okay to say that you "engage" your students, saying that your classroom is "student-centered," whether true or not, simply makes you like every

Teaching Philosophy

My basic goal as an English instructor—based upon a philosophy I call "meta-pedagogy"—has been to make students aware of the educational process itself. Students are encouraged to become active participants in the construction of course syllabi, organization of class activities, and the conveyance of knowledge. They are encouraged to consider the implications of educational policy making and pedagogical presentation so that they might become more critical of the practices that affect their own acquisition and use of knowledge. I have focused on helping them to strengthen their convictions and stressed the importance of articulating those convictions in a variety of settings.

My experience has taught me that students often perceive educators not as people working to help them, but as obstacles or stepping-stones between them and their futures. I've come to realize that such (erroneous) perceptions are partially the result of their detachment from or nonparticipation in the educational system. Most students go to class, take their tests, complete their core requirements, and fill out their evaluations because they are asked to do so but not because they understand the reasons for doing so. However interested they may be in knowing those reasons, they are often conditioned not to ask about them, not to question the purpose or efficacy of traditional or nontraditional pedagogical methods. I have been impressed by the positive reactions of students once they are comfortable enough to ask these "forbidden" questions. For example, the first question I tend to be asked by writing students is "Why do we have to take these classes?" Several years ago, my response was typical of the unsatisfactory answers that are usually given: "Because every job requires written communication skills, etc." Now I assign interview papers that each student must complete. The student must arrange for an interview with a person in her prospective field (a dean, employee, professor, etc.). She must explain to the interviewee the class she is taking, and then she must question how it will be useful down the road. Without exception, students return to class after the interview more determined to work and appreciative of the concrete answers they've discovered.

I have embraced an interdisciplinary, multi-media approach to teaching in order to stress the connections between fields of knowledge that students often perceive to be unrelated. For example, in "Introduction to Shakespeare," we move from an in-depth examination of each play to musical and artistic reconstructions of Shakespearean drama such as Mendelssohn's Overture to *A Midsummer Night's Dream* and Henry Fuseli's painting of the same title. My classes integrate music, film and television clips, and trips to local art collections and playhouses to stress the complex pervasiveness of ideology and the exciting inter-connections between cultural media. Students begin to see knowledge as dynamic and alive, not fixed and static.

In conclusion, I admit that my greatest fear about meta-teaching is that I will be unable to maintain enough authority to conduct an effective course. After all, my courses teach students how to be critical even of me. I've learned that the fear is unnecessary. By focusing students on the learning process, I help them to understand the highly complex factors that influence my assessment of their performances. They begin to feel as though they can understand and control these factors as well. Grades are less frightening as a result. They become markers on a quite accessible pathway to improvement and success. As a teacher, I have tried to empower students while maintaining rigorous standards of excellence.

Figure 6.6 Sample teaching philosophy statement

other teacher on the market. Indeed one advantage of focusing on a single and personal pedagogical ideal (such as "meta-pedagogy"), rather than simply describing your teaching style, is that the former approach allows you to use terms like "critical thinking," "multicultural," and "interdisciplinary" in the service of a relatively original approach. Such cliché terms and concepts only become a problem when they are presented as the philosophy in and of themselves. One does not do anything very special by attempting to foster critical thinking, which after all, is the job of every teacher. But one might catch our attention by explaining to us *how* students can be encouraged to think critically.

Above all else, your job in the teaching philosophy statement is to show that you are a person of conviction, a thinking educator who can transform a theory or principle into a practical tool for improving people's minds.

Student Evaluations Summary

Despite every negative thing I said and meant above in reference to computerized student evaluations, they represent one of the few portable indicators of teaching effectiveness available to us. And despite the fact that I put little stock in their ability to measure teaching performance, I admit to being skeptical about job candidates who fail to provide their averages in teaching portfolios. While average or slightly below/above average scores on the evaluations signify next to nothing, one might infer a great deal from consistently horrendous or incredibly high averages. Since one assumes that any sensible person would be sure to include excellent scores, the absence of *any* information whatsoever raises some red flags. Since I am not recommending that you list poor or below-average numbers, my advice is that you do everything in your power to make sure your averages are not poor or below average. Should you experience problems the first few times you teach, like most young teachers do, you should use every resource available to you to improve your performance; once you've improved, you can list the evaluation results from, say, the last 3 or 4 years. In short, these scores matter, whether they should or not. (If your university has no equivalent of computerized evaluations, you should say so clearly.)

Now there are two ways for you to share your numbers. The first is to list them next to all of the courses you've taught, as in figure 6.7.

Such an approach would allow readers to take in everything at once and also to note your improvement over time (if you have improved

Student Evaluations

Scale from 1.0 to 7.0, 7.0 being highest

	Course	*Instructor*	
SU 00	English 202B	Not available until October	
SP 00	English 129	6.32	6.52
FA 99	English 221	6.11	6.52
	English 221	5.94	6.55
SP 99	English 202B	6.11	6.50
FA 98	English 202B	6.24	6.47
SU 98	English 202D	5.75	6.19
SP 98	English 15	5.62	6.29
FA 97	English 202D	5.30	5.85
	English 202D	5.67	5.60
SP 97	English 15	5.59	6.18
FA 96	English 15	5.92	6.38
SU 96	English 15	5.90	6.65
SP 96	English 15	5.45	6.10
FA 95	English 15	5.10	5.60

Figure 6.7 Sample computerized evaluations report

drastically, you might consider listing the numbers in descending order from the most recently taught course). Make sure that you offer some kind of key for translating the numbers. For example, you should explain that teaching performance is ranked on a scale of 1 to 7 at your university, 1 being the worst score and 7 the best. If your university provides departmental averages that you happen to exceed, be sure to share this information. In the second system, you would combine your numerical scores and students' written comments, as in figure 6.8.

The advantage of this option is that it allows you to paint a more comprehensive picture of a particular class and perhaps to emphasize that substantive student feedback is as important to you as the numbers.

Written comments should be extracted from the evaluations and transferred verbatim (grammar mistakes and all) to the portfolio. Select four or five comments for each class that reveal something you want your readers to know. Do not include any negative comments; honesty is not the best policy in this case. But you should also avoid including positive evaluations that are either vague ("Mrs. Johnson is great") or that are nonteaching related ("Mrs. Johnson's awesome 'cause she brought us candy"). Instead select comments similar to those in figure 6.8, which convey what students have learned in the class or how the teacher's methods helped them to learn. As a new Ph.D., you should include in your portfolio no more than two or three pages of written evaluations.

English 267.01 "History of God" (Fall 2003)
40 Students
Overall Mean: 9.5 out of 10
Overall Median: 10.0

Student Comments
"I liked the progression of literary works from Job to Philip Pullman because we were able to see the formation of the concepts."

"Stereotypes were abolished, new perceptions formed, all due to the effects of this class and the Professor."

"I really liked the class and the way it made me question a lot of my beliefs and ideas that I had taken for granted or never really thought about before."

"I learned more about biblical/satanic history than in my 14 years of religious education classes. It was stimulating and interesting. One of the better, or best, professors I've had in my four years."

"The professor was always very energetic in his teaching style, often passing that enthusiasm onto his students. He also made himself very available to his students and is always responsive to our ideas and thoughts."

"Easily the most interesting class I ever took."

English 221 "Survey of British Literature" (Spring 2003)
24 Students
Overall Mean: 9.6 of 10
Overall Median: 10.0

Student Comments
"The professor was able to take some of the most difficult literature there is and not only present it in an understandable and interesting manner, but to relate it to all sorts of things the class could understand."

"I have to say that this class is the first one I've really learned anything from in terms of writing ability. I credit that to the professor and the simple fact that he wouldn't give an A to anyone who could write a sentence. Although I don't think that my grade will be as high as in other courses, I am incredibly pleased with the knowledge that I gained in this course."

"His enthusiasm exceeds that of any professor I've had."

"Sitting in a circle and discussing important issues, we read great literature, and the professors showed a lot of passion for what he was teaching. I learned more in this class than I have in any other class in the 4 years I've been in college."

"The most positive aspect of the course is how clear it is that the professor cared greatly for each of us and our education. It's obvious that he puts us before himself in many ways."

Figure 6.8 Sample computerized evaluations report with comments

Syllabi

Next provide abridged versions of your two most impressive syllabi. If you have taught upper-level classes as well as introductory ones, provide one for each type. In addition, provide your potential employers with one syllabus for a course you plan to teach if they hire you.

Look up their course catalogue and requirements, and tailor the course to their program. Make sure that the course fulfills the specific needs underlying the job search. For example, if you happen to be a Racine scholar answering a job ad for a seventeenth-century French drama scholar, you should work up a syllabus either for an upper-level class on Racine or a survey of seventeenth-century drama.

Since you should try to keep each syllabus fairly concise (no more than two pages), cut all personal information such as your office hour times, your phone number, and your policies regarding such matters as plagiarism, attendance, and lateness. The latter is especially important since it's not worth risking that you might offend someone because your policies happen to be more rigid or relaxed than their own. If your audience cares about such things, they will question you about them during an interview. The most important information to include on each syllabus is the course title, course description, required texts, required assignments, and the daily class schedule (see the examples in the Appendix). Remember that your syllabi will convey much about your level of organization, your commitment to learning, and your vision and understanding of a particular subject. Rhetorical flair is not the goal here. While each syllabus should be "sexy" enough to win over the students as early as day one, it also should reveal a teacher whose expectations for hard work are reasonably high.

Letters of Observation

By showing how much work must go into the portfolio, this discussion has hopefully convinced you that you should begin preparing your own one long before you go on the market. One section of the portfolio you simply cannot save for the last minute will include observations of your teaching from faculty members and, if you are ambitious, your former students. Since you should probably insert no more than three letters into the portfolio, it's important to be strategic about whose letters you include. Unlike the recommendation letters in your dossier, you must have regular access to these letters so make sure that the writers are comfortable allowing you to read their observations. Regarding which faculty members you should ask to write, make it a point to be observed by your major advisor at least once before beginning your final year of graduate work. Remind her that the letter she writes about your teaching can also be incorporated into her general letter of recommendation, which will save her time later. Your advisor's observations will mean more if they pertain to a class related to her area of specialization. At least one other letter should

Abridged Teaching Portfolio*
Sergio Valenti

Contents

*Longer versions of this portfolio are available upon request.

Figure 6.9 Sample portfolio contents page

come from a faculty member, though the writer need not be a member of your dissertation committee. Simply select someone who you believe is likely to understand your approach to teaching and will be willing to write you a strong letter of support. Finally, try to obtain a letter from a former student. We all know excellent students with whom we stay in touch after our official time together has ended. Allowing such a student the ability to say "no," simply ask whether she would be willing to write you a strong letter of support for your file. Be sure to inform the student that because the letter will go into your portfolio, you will be able to read it. If the student is agreeable, consider coaching her gently on what such letters tend to look like, even provide a sample if you have one, but do not dictate the form or content of the letter. My advice would be to collect several such letters over the years so that you might include in your portfolio the most impressive one.

Mechanics

Work up a "Contents" page for your portfolio (see figure 6.9), print out the forms on both sides of the page, and staple everything together neatly.

You now have a concise and useful record of your teaching life, which you can use immediately for the job market and later on when being considered for teaching awards, tenure, or promotion. To see what a complete portfolio looks like, consult the appendix (pp. 295–300).

The Research-Teaching Connection

A common question asked by interviewers for tenure-track jobs is "In what ways has your research affected your teaching." People generally are of two minds about how to answer the question. On the one hand, most of us recognize that our research is much too specialized and focused to bring into the undergraduate classroom without serious dilution. This is the wrong answer for a job interview, however. On the other hand, when we actually think about it, we recognize the myriad ways that teaching and research are interconnected, even to the point where it becomes seemingly impossible to sum up the nature of these connections. By coming to terms early in your career with the important relationship between teaching and research, you'll make yourself a more efficient and effective practitioner of both arts.

The simplest point to stress is that your research training is precisely what defines your ethos in the classroom. Were you not a highly trained expert (or expert-to-be) on a particular subject, you would never have been awarded the right to teach at the college level. Now what does this mean, exactly? Notice that when we are asked about our research, we think automatically about our specialized area or current topic of interest. We might also usefully consider our ever-expanding knowledge of the research process in our disciplines. That is, training toward the Ph.D. makes one an expert in English or Art History *as well as* atheism in Christopher Marlowe or eroticism in Praxitiles' sculpture. With every new book you read, every seminar paper you write, every grammatical principle you master, or every electronic database you learn to use, you are practicing what it is you've been hired to preach. Imagining teaching apart from research, in other words, should be nearly impossible to do.

Where things become tricky is where the subject matter becomes increasingly specialized. But here too teaching and research are inextricably linked and their influence is reciprocal. As a Renaissance specialist, I am somewhat embarrassed to admit that I spent most of my graduate career believing Edmund Spenser's *The Faerie Queene* to be a tedious work. Then I was asked to teach it. Forced to say something interesting and insightful about the epic, I read it more carefully, until I began to notice how meticulously constructed every stanza of the massive work was—how functional and systematic every single preposition was. The excitement I experienced as a result of my own discoveries as a reader was compounded by the knowledge I began to accumulate as a result of turning my newfound energy toward the secondary literature on the poem. Of course, my enthusiasm also spilled over into the classroom, where suddenly *The Faerie Queene* had become a living, breathing work of art for my students. At a certain point, one loses the ability to distinguish between teaching and research; the dualism is a false one. Teaching at its best is a constant process of researching materials and ideas. Shared research is teaching.

The key is learning while you are still a graduate student to take advantage of the reciprocal relationship between teaching and the research you do as a student, an examiner, and a dissertator. In a pad or computer file labeled "Future Projects," you should begin keeping track of all your emerging ideas either for courses you would like to teach or articles/books you would like to write. You will find while reading materials for your own seminars, for example, that certain books would work perfectly in a particular class you plan to teach in the coming years. Writing your ideas down now will make the process of composing the syllabus much easier later on. More often, your discovery of certain ideas in the classroom will feed numerous research projects over the course of your career; in fact, it's probably safe to say that most research projects grow out of discoveries related to our engagement of materials for class. Since you will always already be busy at work on at least one other research project, however, it's extremely important that you keep a record of ideas; otherwise, you will simply forget them.

Finally, be careful to save and organize all of your teaching-related materials. Not only will they be helpful on those occasions when it becomes necessary to arrange or revise your teaching portfolio, but they also will serve as a major reference library for a variety of future teaching- and research-related activities. We have already discussed the value of building files of information and materials gathered during the course-work and examination stages of your career. Equally important will be those files classified under "Courses Taught." Since you are very likely to teach the same or similar material more than once and perhaps throughout your career, careful record keeping will minimize the work you'll need to do each new semester.

As this chapter would suggest, I see little point in denying that teaching *can be* one of the most stressful and frustrating jobs imaginable. Especially if we fail to recognize the importance of the work we do in the classroom or if we dichotomize too severely our classroom- and research-related activities, teaching can become merely a burden—something to be dreaded. If we invest fully in our teaching, however, and learn how to minimize obstacles to success in the college classroom, most of us will discover the unparalleled joy and satisfaction that comes from helping others to know themselves.

Chapter 7

Exams

Perhaps no requirement for the Ph.D. is more dreaded, more misunderstood, and more badly mismanaged than comprehensive examinations or "comps". Whereas preparation for the master's examination tends to be relatively self-explanatory, since students usually are asked to respond to a standard or fixed list of works (see pp. 41–44), Ph.D. comps often cause students trouble because they represent the first stage in one's graduate career where content is determined largely by the student, rather than imposed from on high. Because you will be responsible for selecting the subject categories on which you will be tested, constructing (with help from your committee members) appropriate reading lists, and managing the long period of study prior to the actual examinations, there simply are more things that can go wrong unless you are systematic, focused, and well-organized. While this chapter explores ideas and strategies also relevant to MA-exam preparation and testing, it focuses primarily on practical solutions for the problems commonly faced at the comps stage of one's career; its subjects include:

- Emphasizing process over product
- Selecting the right exam areas
- Composing reading lists
- Streamlining for the dissertation
- Reading and note-taking
- Understanding grades and comments
- Preparing for the oral defense

Despite the difficulties they present, comprehensive examinations can be productively regarded as the transitional step from the relative dependence of the graduate student in course work to the coveted autonomy of the advanced scholar. For all intents and purposes, no one after comprehensive examinations will ever again decide for you what you will read, research, or write. In what follows, then, I discuss how to handle an experience that should be at least as liberating and enjoyable as it is daunting.

THE PURPOSE OF THE COMPREHENSIVE EXAMINATIONS

No two exam systems are alike. At the University of North Carolina at Chapel Hill, history Ph.D.s are tested on a single list of about a hundred works "critical to an advanced understanding of a particular field." At the University of Washington, literature students are tested by written and oral exams on three separate lists covering a major period, a major literary genre, and one specialized field such as rhetoric or theory. At the University of Connecticut, English students first take a three-part general examination designed to broaden and ensure their "teaching knowledge," and they follow it up several months later with a "specialist examination" geared more directly toward the dissertation. Humanities students at Yale test orally on nine officially recognized fields out of a possible twenty-five. In all of these cases, Ph.D. examinations serve to define *in terms of content* one's expertise in a relatively specialized body of knowledge.

It's rather easy to see that all of these exams are flawed in one way or another. UNC's specification of 100 works is about as random a number as any other, and the phrase, "a particular field," is notably vague. Washington's system, which is a variation on the most common type of Ph.D. exam, defines "specialization" in a fairly narrow manner. UConn's two-part examination makes it extremely difficult for students to complete their examinations in a reasonable amount of time. Yale's system is highly prescriptive, suggesting that expertise in a particular area can be defined in canonical terms; further, the absence of a written exam means that students will be tested less on what they know than on how well they can stand up under interrogation.

Now it would be unfair of us to deny that if all of these exams seem a bit flawed, they do so in large part because the concept of a standard examination is, by its very nature, a bit flawed. By the time you finish your reading for examinations, you will understand perfectly well that no examination will be able to test you effectively on how much you know. Nor will any examination be able to define very clearly what it is that you should know. Most important, no examination will be able to indicate with any degree of accuracy how well you will perform as a teacher or scholar at the professorial level. Surely most people in academe understand these realities. So what's the point?

Three Variations on a Practical Question

The Idealistic Response

Comprehensive examinations ensure that all Ph.D.s claiming expertise in a particular field possess at least a basic knowledge of central texts and ideas in that field. A common response, but the assertion is so problematic that one hardly knows where to begin. First, it implies that exams cover a standard body of material that is disproved even by a glance at the few examples I've provided above. Second, it implies that "central texts and ideas" can be covered by a list of a hundred or so works. Finally, it implies that (1) we can differentiate central and peripheral texts and ideas, and (2) that only the ones arbitrarily labeled "central" are worthy of our attention.

The Cynical Response

All of these exams were taken—in one form or another—by the professors who came before us; therefore, we must endure the same sort of "boot camp" training suffered by our advisors. We would be dishonest if we tried to deny that this type of mentality sometimes does apply in academe, but in the case of exams, it requires that we also deny altogether the numerous long-term benefits of the Ph.D. examination process.

The Realistic Response

The preparation for examinations—rather than the examinations themselves—forces us to undergo a rigorous period of study designed to enhance our knowledge of undeniably complex fields and subjects. While such a claim requires that we define "knowledge" in a somewhat vague manner, most Ph.D.s who have survived comprehensives would probably agree with the claim's basic accuracy.

The key to making your exams useful is your ability to stay focused on the process of studying rather than on the product (i.e., the graded written and/or oral examination) in which your study culminates. The sanity of this approach is indicated by the relative uselessness of that product, once you've been permitted to move on to the next stage of your career. Your examination grade will not appear on your official transcripts. More important, no one likely to hire you will care even the slightest bit about how well you happened to perform in your exams (though, in an interview, they may ask you which subjects you chose to focus on). Potential employers will simply assume that you studied hard for your exams, that you increased your knowledge of your major field as a result of that hard work, and that your exams have informed directly both your dissertation research and your course content. Since this will be their assumption, why not make it work for you? In much of what follows, I highlight various strategies for using examinations as a starting point for dissertation research and advanced teaching.

THE PAINS (AND PLEASURES) OF THE EXAM EXPERIENCE

I'll be honest with you: the only time in my graduate career that I seriously considered an alternative career came while I was preparing for comps. After years of unwavering certainty about my life and career goals, I suddenly found myself questioning not only my own abilities, but the very soundness of the whole enterprise. Now it makes a certain amount of sense that those doubts arose when they did; in a way, what I call "comps burnout," a phenomenon experienced by about half of the Ph.D.s I know, represents the graduate school equivalent of a midlife crisis. Somewhere in the transition from student-status to professor-status, a transition marked by the newfound independence of the soon-to-be ABD, many people tend to lose their way. On the one hand, this may not be such a big deal; every graduate student probably should, at some point, face the hard questions about where he's going. On the other hand, self-interrogation can be merely destructive if one doesn't keep things in perspective. The Ph.D. examinations seem particularly dangerous in this regard because—as an anomaly in academic life—they allow one too easily to forget what one loves about academe.

Common Problem #1: Desert Island Syndrome

Simply put, the most common problem is the isolation. Month after month after month in your study, alone, reading book after book after book. "I'll skip the beer tonight because I have too much reading to do." "I won't go in to campus for that talk because I have too much reading to do." "I shouldn't go for my run today because I have too much reading to do." The furniture begins to collect dust. At a certain point, you lose the ability to determine when it is day and when it is night. The hair on your head hasn't been combed in days. Why bother? No one will see you. You'll be at home, alone, reading book after book after book.

The first key to surviving examinations and staying healthy is maintaining your social ties—from family and friends to advisors and classmates. Set up a meeting with each of your advisors every few weeks so that you can discuss the material you've been reading. Find a friend who's also in the process of reading for comps and spend one day a week reading together in the library. Give yourself a deadline for completing your day's work and then go for that beer. And, by all means, stick to your exercise program. By forcing yourself, in other words, to set up and adhere to a schedule of events, you can establish structure and order where there is otherwise only time and space.

Common Problem #2: The Anticlimax

The second problem is the letdown that often follows completion of the exams (which also is experienced at the MA-exam stage). By the time you walk into the testing room, you'll have more knowledge at your fingertips than you ever imagined possible. You'll be dying to show it all off. Then you'll look at the question, which will ask you to do something like compare two texts' different treatments of some obscure issue. Wait a minute, you'll say, I just read several hundred books and articles, and this question wants me to talk about *two* of them? You'll easily be able to answer the question, but you'll leave the testing room feeling dissatisfied by the experience. A Ph.D.-level training for a high school-level test. And you probably won't get a very high grade on it either.

The only real solution to this problem will be your ability to anticipate it and to stay focused on what's really important, which is the process of learning the material in the first place, as we discussed above. The simple fact is that there are very few exam questions that could satisfactorily measure the extent of a Ph.D.'s expertise in his specialized area. Rest assured that your dissertation will serve this purpose more than adequately. For now, make it a point to read all of the material and be prepared for anything, but know that the exam will most likely leave you feeling somewhat cold and perhaps just a little bitter (see more below on "Taking the Test", pp. 160–63).

A Strange Love? or: How I Learned to Stop Whining and Love the Exams

What's ironic about the fact that so many people wind up hating their exams so much is that the comps period *could be* the best time in one's graduate career if one were to approach them sanely and systematically. Think about this: your advisor tells you to go away for four or five months so that you can read a hundred or more books. You don't have to go to class anymore. You don't have to write seminar papers.

You don't even have to leave your apartment. You just have to read. What most Ph.D.s realize long after comps have ended is how badly they long for the freedom to read and learn that they were granted back then. Take my advice: be sure to exercise whatever precautions are necessary for you to avoid becoming lonely or embittered, but remember to enjoy the freedom you've been granted. When you begin your preparations for comps, you will know a lot about your field. When you turn in your exams half a year later, you will be an expert in your field. This sort of intellectual growth is precisely what you came for.

THE EXAM PROCESS

Preliminary Steps

Ideally, your planning for exams should begin at least as early as the first semester of your fourth year in graduate school (see pp. 47–48). Unnecessary delays can result from poor planning, so make it a point to do your homework early. By the time you begin preparing for exams, your advisory committee will begin to perform—probably for the first time in your academic career—a significant role in your life. Once you've chosen the members of your committee (see pp. 45–47), you'll want to take the following immediate steps:

1. Meet with your major advisor to assess and discuss your long-term plans for completing your degree. Because you'll want to be as informed as possible for this meeting, be sure to familiarize yourself with the department's policies and practices regarding Ph.D. examinations. Explain first to your advisor when you would prefer to complete your exams—preferably no later than the midpoint of the fifth year. Stress the importance to you of staying on track for graduation, and seek whatever advice about doing so your advisor may have to offer.
2. Establish with your advisor the categories or subject areas on which you will be tested. Remember that your exams are more likely to be useful to you if they allow you to prepare for your dissertation; with this idea in mind, you should make it clear upfront those areas *you* would prefer to study; if you show up at your advisor's office without a clue, your exam areas will likely be dictated from above. Obviously, you must be willing to entertain your advisor's objections or suggested changes to your plan. The more persuasive a case you can make for your own choices, the

greater the likelihood you'll leave your advisor's office as a happy person.

3. Seek to understand your advisor's philosophy or perspective of the Ph.D. examination process. Does he see the exams mainly as the culmination of a student's course-work? As a preparation for the dissertation? As a preparation for teaching? What does he believe are the differences between an excellent and a poor examination? How does he understand his role vis-à-vis the other members of the committee? You would be wise to discuss the same issues with your other advisors. You can avoid the sorts of problems I discussed above by making sure you know what is expected of you from each member of your committee.
4. Finally, ask whether it would be okay for you to see an example of the type of exam your advisor has given in the past. While you'll want to collect as many old exams as you can get your hands on (ask friends and consult the graduate director, who in many cases, will keep an exam file for students to evaluate), the most useful examples would obviously be those created by your advisor. If he does not keep a file of old exams, track down one of his older advisees and ask if you can take a look at his exams. Though you should be prepared for anything once you enter that testing room, some scientific speculation about what you'll encounter there can only help you to be prepared.

Composing Your Lists

Assuming that your department does not require adherence to a fixed list of works for each subject area, you'll want to begin constructing your own lists as soon as possible after the meeting with your advisor. It may make the most amount of sense to do this over the winter break and then seek approval of the lists in the early spring. In any case, you should begin as always by collecting others' materials. Usually at least one "older" person in your program will have tested on each of the subject areas you've specified for your own exam; ask them for their lists. You won't benefit from copying them without change, but you will get a sense of what such lists look like and also of the sort of lists the department has approved in the past.

Keep in mind two questions while composing your exam lists: which primary texts will I likely be expected to teach once hired as an assistant professor? And which texts will I need to read in order to complete my dissertation prospectus shortly after I finish exams? It goes without saying that you'll benefit greatly by combining your examination and dissertation research as much as possible.

However else you choose to organize the works on your lists, you should be sure to include the author's name and the exact title of the text, and to arrange them in some logical, easily comprehensible manner. You may find that chronological organization works better for you than generic or thematic organization, which is fine. As always, though, you should follow some type of system. Occasionally, it will be sufficient simply to arrange works according to the subject categories on which you will be tested (perhaps with a few minor variations), which is what I did for my own exams. Talk to your advisors and stick with what works for you.

In retrospect, one significant problem I find with my own lists—and something I would correct if I had it all to do over again—is the absence of complete bibliographic information for each work. Such a level of detail will rarely be required by a Ph.D. candidate's committee or department, but it would behoove future dissertators to spend the extra time it takes to provide the complete information, for two specific reasons. First, many, if not most, of the works included on your list will play some part in your dissertation; by providing more information at the comps stage, you'll be in a position simply to cut and paste material later on, when you're likely to be much busier. Second, a detailed list will place your advisors in a better position to suggest the most appropriate editions for you to study. It makes little sense to read the Oxford Classics edition of a text for your exams if your committee is going to require you to use the Revels edition for your dissertation.

Once you've drafted your lists, seek approval first from your major advisor and then distribute copies to all of your committee members. Plan to meet with each one of them a few days later to discuss recommended additions to and deletions from your lists. Keep in mind that your lists will very likely grow larger as a result of these meetings; only rarely will someone recommend that you drop a text from your lists. Only after each member of the advisory committee has approved your lists should you seek official approval of them by the graduate director. Once you get it, it's time to schedule the exams and start reading.

Scheduling Exams

Most programs allow students to space out individual exams within a reasonable frame of time. For example, most programs allow students to take exams consisting of three parts either three days in a row or over a period of approximately three weeks, with a week-long break in between each one. While many students choose to spread out the exams as much as possible, you should not rule out the idea of testing two or three days in a row (a compromise might be to test on Friday, Monday, and Tuesday, which allows you to rest over the weekend). The disadvantages of trying to knock out all of the exams at one time include forcing yourself to study for all of the exams prior to taking the first one, and the sheer exhaustion you will feel as a result of doing so. On the other hand, you are bound to be exhausted on the final day of exams even if you spread them out over several weeks, and you may only be prolonging your stress and exhaustion by taking more time to get them done. Think about your own work and study habits, seek the advice of your committee members and more experienced graduate students, and make an informed decision.

When scheduling your exams, be sure to arrange for the testing conditions most amenable to your own needs and work habits, provided that your department allows you to do so. If you suffer from claustrophobia, ask for a room with windows. If you have a hard time composing on a computer, seek permission to write your exams (some departments that require typed exams will allow students to write them out first and then type them verbatim into a computer). You might not be allowed to use your own computer, in which case be sure you know how the one you will use works. Figure out those materials that you will be allowed to bring into the testing room (books, notebooks, etc.). Finally, ask permission to see the room prior to your exams.

The Reading Process

How do you plan to proceed? Will you tackle one subject area at a time? Will you read works in chronological order? Every time you think you've worked out the perfect system, it's time to revisit the same old questions. Rather than worrying too much about the right order in which to read everything, I recommend focusing on the time frame in which you plan to read everything. The greatest danger is that if you stray too far from this plan, you'll be more likely to forget much of what you've read early on. Remember that most people read more carefully and even leisurely in the first few months of study and speed up the process later, once they've discovered how little time remains before the big day. With this in mind, work to establish a reasonable time period for all of your reading and pace yourself as sensibly as possible.

Since you'll never forget the works you read for your Ph.D. exams, and since they will all be eternally associated with one another in your mind, I recommend keeping all of your exam notes in one place—either a large, sturdy notebook or a computer file. Excessively detailed notes won't be necessary or even useful in most cases. Keep in mind that what you'll need to remember once your exam begins is actually pretty basic information: for primary works (whether plays, pamphlets, or paintings), the basic form, style, and distinguishing features, as well as a sense of the context in which it was created; for secondary works, the argument, methodology, contribution, and, in some cases, the context in which it was created. You might also wish to record observations about the connections between the various works on your lists. To this end, you may wish to create index cards for each work on your lists, which will be useful study aids just prior to your exams. On the front of the card, jot down the author/artist, date, and title of the work; on the back, write down four or five important features of the work. Now you can practice memorizing what you need to know about each one. While more detail typically won't be necessary, you should also keep in mind that the notes you take may prove useful later on at the dissertation stage, so be sure to write down and mark clearly any ideas or passages, however minor, that you anticipate returning to after exams.

As a practical matter, you should always consider the work you've just read in relation to the category or subject area in which you've classified it. The questions your committee will ask you are very likely to engage the relationship of the texts you've chosen to study and the larger contexts in which you've chosen to organize them. That is, how one reads Diderot in the context of the Enlightenment will differ from how one reads him in the context of the history of encyclopedias. How one reads Picasso in the context of Cubism will be much narrower than how one reads him in the context of twentieth-century painting. While you should anticipate questions that touch on any or all of the possible ways of viewing a particular work, you can estimate—and figure out from old exams—those types of questions you are most likely to be asked.

Finally, make it a point to enjoy the reading period. This may be the last time, after all, that you find yourself in the enviable position of being allotted the time to read so broad a body of material for such personally and immediately useful purposes. Don't allow the anxiety you feel about the upcoming exams to overshadow the significant pleasures you can reap from a heightened understanding of the subject matter you value most deeply. Stay focused on the process, and let the product take care of itself.

The Form of the Test

If you do your homework, it's unlikely you'll be surprised by the questions you encounter on exam day. Figure 7.1 presents a typical form for one part of an English Renaissance literature examination.

Notice that the examiner allows the student to choose among at least two questions for each section. Only in very rare cases would a

---

PhD Day-One Exam in Renaissance Literature

Please answer one question from Section I and one from Section II. Make sure that you write about at least six different texts in the course of the two essays.

Section I Questions with a Broad Scope: Write on one of the following topics.

1. Partly because the terms "homosexual," "heterosexual," and "bisexual" had no meaning in the Renaissance, scholars have emphasized the period's relatively fluid conceptions of sexual desire and orientation. Choosing at least three representative texts upon which to ground your argument, discuss how Renaissance authors represented the ever-problematic relationship between sexual desire and social orderliness. Define your terms clearly, and try to address a broader chronological range of texts from the sixteenth and seventeenth centuries.
2. In a recent study entitled *Shakespeare's Tribe* (2002), Jeffrey Knapp urges readers of Renaissance texts to revisit their most basic assumptions about English nationalism in the sixteenth and seventeenth centuries. Whereas critics and historians often discuss the emergence of nationalism in the period (Richard Helgerson's "concerted generational project," for example), Knapp argues that Christian/religious identifications remained throughout the period much more important than nationalist ones. According to Knapp, even in so seemingly jingoistic a group of texts as Shakespeare's English history plays, England's relations with foreign countries are defined more by religious matters than secular, political ones. This sort of break between the religious and political realms may or may not be useful. Focusing on at least 2 primary texts from your list and one other canonical Renaissance text, discuss how religious and nationalist identities and identifications tended to operate in Early Modern literature; were stereotypes of the Dutch, for example, based more on religious or other political factors?

Section II Questions with a Textual Focus: Write on one of the following topics.

1. If the Lutheran notion of "grace by faith alone" highlighted the sovereignty of the individual, and Calvin's doctrine of double predestination highlighted the sovereignty of the godhead, then Arminianism highlighted the fact that the individual *and* God could be sovereign at the same time. Discuss the problem of "free will," and the techniques for representing it, in relation to *Paradise Lost* and two earlier works.
2. Renaissance authors inherit an Aristotelian poetics that tends to privilege tragedy over comedy, but they also demonstrate a fairly revolutionary willingness to mix "clowns and kings." Focusing on Shakespeare and two other writers from your list, discuss how Renaissance comedy, like tragedy, serves to instruct or "fashion" readers/playgoers.

Figure 7.1 Sample exam form

Ph.D. student be forced to answer a single question (oral exams allow less choice, of course, but they also permit various opportunities for clarification and qualification of issues). As in the case of any examination, you should not let your eyes read ahead of your mind. What I mean by this is that an overly quick scan of the questions may only convince you that you don't know anything, adding to your already-considerable anxiety, and paralyzing you where you sit. When you receive the exam, take a deep breath, study each question carefully and slowly; think about how terms are defined and might be redefined; consider what sorts of examples you might invoke to answer the question; now take another deep breath and demonstrate what you know.

Taking the Test

You will learn more from looking at old exams, and discussing with your advisors what's expected of you, than you will from trying to define in abstract terms what constitutes a good exam. The simple fact is that different committees will understand the matter in different ways, and departmental policies influence directly how a particular exam is to be evaluated. For this reason, only a few general comments are in order here, and they apply in the cases of both oral and written exams.

Make Sure You Answer the Question

Does this advice seem too obvious? If so, you'd probably be surprised by how common it is for students to "spin" questions in every direction but the one preferred by the examiners. The problem occurs for one of two closely-related reasons: either the student feels intimidated by the difficulty of the question and so attempts to make his task easier by altering or even simplifying it; or the student attempts to demonstrate his expertise in the given area by moving beyond the range of issues implicitly relevant in the specific question. In the latter case, no one should ever fault a student for trying to make a question his own, so long as he manages at some point to address what the committee most wants to know. In the former case, the first thing to recognize is that Ph.D. examination questions often are hard, and there's no shame in acknowledging this fact. If you're having a difficult time understanding what's wanted of you, you have the right to request a clarification. If no opportunity for a clarification exists, begin your response by articulating exactly how you interpret the question (e.g., "My understanding is that by 'post-modernism,' you mean . . .") and then proceed. You may wish to entertain alternative possible responses just to be sure you're covering all of the bases (e.g., "If on the other hand, you mean by 'post-modernism' Y, then I would argue that . . ."). If the problem you're experiencing has less to do with basic comprehension and more to do with a feeling that you simply don't know the answer, calm yourself down and call to mind one basic rule: at this level of learning, there's almost never just one right answer. As long as you make a sincere attempt to answer the question you've been asked, it will be acceptable, nine times out of ten, for you to push the discussion in a relevant direction in which you feel comfortable moving.

Complicate Questions Rather Than Simplifying Them

You're very likely to encounter a number of questions phrased in either/or terms. For example, consider the following question, which appeared in one of my Ph.D. exams:

> By 1990, reviewers were proclaiming that the New Historicism was dead, but as one reviewer also noted, its prime progenitor refused to lie down. Looking at the work of Stephen Greenblatt, and of two other scholars who have been termed "New Historicists," explain whether or not you think that the proclamation of death was—or is—premature.

In simpler terms, we might say that the question asks whether or not New Historicism is dead. Although either a simple "yes" or "no" answer would amount to an oversimplification of ridiculous proportions—the question is a trap, in other words—many test-takers would be willing to oversimplify it for one of the following two reasons: because they feel like a "yes" or "no" response is necessary to answer the actual question, or because they feel that a more subtle response would suggest a lack of conviction.

Don't fall prey to the either/or trap. Instead, without insulting the examiners, interrogate the question itself. First of all, what do we really mean by "dead"? Are we asking if the New Historicism still has a place in journals, intellectual conversations, and graduate courses? Of course, we know that it does. Or are we asking whether or not its basic tenets are still persuasive? What are its basic tenets? Are Greenblatt's Historicist tenets the same as those of "two other scholars"? And what do we mean by "New Historicism" anyway? Are we referring to a perspective of how we should understand historical artifacts? Are we referring to particular critical methodology that begins with a historical anecdote, à la Foucault, and then draws from it more general conclusions?

I do not wish to suggest that you raise all of these questions on the examination itself, since you may only succeed in irritating your examiners by doing so. But you need to come to terms with how you plan to approach a complex question dressed as a simple one. Here's how I chose to answer it:

> If we mean by the term, "New Historicism," a theoretical approach to reading texts that solves all of the problems of the so-called "old history," then we would have to say "yes", the movement is dead. Numerous observers have pointed out the flaws in a system that attacks formalism and monolithic power structures but claims, as one of its basic tenets, that subversion is *always* already contained. If on the other hand, we mean by "New Historicism" an approach to reading historical texts that challenges the belief in one indisputable or even knowable reality, that considers texts in relation to the culture that produces them, and that shows an awareness of its own contingency, then we would have to insist that the "New Historicism" is anything but dead.

There are many things that I would change about this response if I could rewrite it now. But the rhetorical solution still seems basically sound to me. I chose to complicate the question rather than oversimplify it, and in the process, managed to articulate my convictions and answer the question I was asked.

Consider the Various Temperaments of Your Committee Members

I remember panicking when I saw the question about the death of New Historicism because on my committee were two self-professed historicists and an individual who regularly criticized the New Historicism in every class he taught. I recognized immediately that an enthusiastic "no" to the question would win me the approbation of the historicists; I also knew it would make me look unsophisticated to the anti-historicist. What to do? By assuming a more objective voice, and defining and contextualizing terms in as clear a manner as possible, you'll avoid stepping on toes. The solution is not to sell out on your convictions but rather to be subtle enough that you can articulate those convictions persuasively to people with very different beliefs and ways of understanding their field.

Finally, Show What You Know

We discussed above the fact that exams test you on a frustratingly small part of what you know. Consider this example:

> Focusing on two artists from each country, discuss the differences and similarities between the German and French expressionists.

Now the question is very specific; its author clearly asks for an in-depth analysis of four artists. Choosing three obviously is not an option. But there is nothing to stop one from establishing a slightly larger context for understanding the "similarities and differences" of German and French expressionism, and your answer is likely to be more persuasive if you recognize the point that two artists alone can in no way represent an entire national movement as complex as "German expressionism." Here's where all of the knowledge you've accumulated can be proudly displayed. Rather than pointing out the characteristics of two French expressionists and then discussing how the works of two Germans match up with them, consider the following approach: offer a history of the two movements—their complex historical and political origins, their underlying aesthetic principles, and their chief proponents—and then use specific works by at least four artists to demonstrate how such global factors translate in local and material terms.

Preparing for the Oral Defense

Should your department require a formal meeting with the advisory committee—which may or may not be called a "defense"—try to view it as an opportunity rather than a burden. One function of an oral exam is to assess how well students can think on their feet since the skills required by an oral exam will also be vital on the job market; therefore, you might effectively approach orals as yet another opportunity to prepare for interviewing. On the local level, think carefully about what you perceive to be the strengths and weaknesses of your exams, and be prepared to discuss both openly and honestly. A common mistake made by students at such meetings is to try to avoid problem areas and controversial points. In order to ensure that your defense goes as well as it can, take the following steps:

Introduce Your Exams

Ask your major advisor for an opportunity to address the committee at the beginning of the meeting. Clarify your reasons for testing in the areas you happened to select, remind committee members of the connections between your exams and your dissertation topic, and seek to articulate your approach to answering the examination questions. Take special care to explain how you might have elaborated or improved upon certain answers had you had more time. Such an approach will show your ability and willingness to critique your own work, and it may go a long way toward convincing committee members that additional criticism would be superfluous.

Be Willing to Say "I Don't Know"

Unless you're absolutely certain that you *should* know, it's okay to be honest about the limitations of your knowledge. Do your best to think critically about the question, offer what you feel is a logical response, and stress that you're still thinking through the issue; you might also invite others' insights on the matter: "Though I'm still wrestling with this problem, I believe that X = Y because of Z. But I would also be interested to know how you understand the problem?" Always remember that a thoughtful response will go over better than a clever, evasive one.

Never Allow Yourself to Become Visibly Angry or Frustrated

Much of the point of any oral examination is to assess a candidate's ability to hold himself together under pressure. An academic who fails to do so, it will be inferred, is likely to be a weak teacher and even more likely to be a poor interviewee when it comes time for seeking a job. Even though you may feel that a particular committee member is pushing you overly hard, it is not appropriate to respond in anything but the most professional manner. In most cases, if you maintain a high level of professionalism, other committee members will recognize your dilemma and come to your rescue. If they see you crumbling, they may only react as other predators do: they may jump all over you. Maintain your poise.

Enjoy the Spotlight

Finally, keep in mind the fact that suddenly, after years of feeling insignificant next to your professors, your work is being taken very seriously and is even the momentary center of attention. Relish the opportunity to show off what you've learned, and be enthusiastic about moving on to the dissertation. The more confident, upbeat, and focused you appear to be, the smoother your defense will go.

Understanding Your Grade

Exam grading systems differ by institution and department. Though in most cases, the relevant categories consist of "Pass" or "Fail", commonly "Low Pass" and "Pass with Honors" are options and, in some cases, regular grades ("A" through "F") are given. The simple fact is that your grade doesn't matter a whit, so long as you scrape by, except in terms of how it might affect you emotionally. Your transcripts won't list the grade, interviewers won't ask about them (if they do, tell them you passed), and your friends won't love you any less should you do poorly on them. To stay focused on the important process of reading, rather than the insignificant product to which your reading has given birth, you'd be better off not caring much about your exam grades at all.

Infinitely more useful would be a series of conversations with each of your committee members about the examinations. Seek to know from them how well they think you handled yourself, how they might have answered questions differently, and most important, how you can sharpen your ideas about matters relevant to your soon-to-be-written dissertation. After you leave their offices, recognize that you are now a much more independent scholar than you've ever been.

From Examinations to the Dissertation

I would classify as a major blunder any student's decision to take Ph.D. exams prior to determining the dissertation topic. Such a blunder would require that the student begin his dissertation research at the prospectus stage rather than six months (or more) earlier, at the exam reading stage. This would also require that the student double or even triple the amount of time he spends writing the prospectus, which could, in turn, lead to months, even a year, of extra time in graduate school. Avoid such problems by figuring out your dissertation topic while still enrolled in course-work, and make it a point to take the following, additional steps to streamline your graduate work:

1. Begin drafting a dissertation bibliography as soon as you figure out your topic. By conducting even the most basic research on the topic and by discussing it with your advisors, you'll learn very quickly those texts that are likely to figure centrally in your dissertation. You can then do your best to make sure that most, if not all, of them appear on your examination lists. Careful reading and solid note-taking at the exam reading stage, then, will help you to knock off two birds with one stone.
2. Create a "notes" file for recording dissertation-related ideas during your exam reading stage. You'll probably be surprised by how often exam readings provoke ideas that can be easily incorporated into the prospectus, the actual dissertation, or both. Cut down on the amount of time you spend writing the prospectus by building its basic contents ahead of time.
3. Stay focused on compositional and rhetorical issues when studying secondary materials for the examinations. Sometimes it is very hard just trying to figure out what a particular historian or critic is saying,

let alone, focusing on *how* he happens to be saying it. Remember, though, that in a few short months, you'll be writing what essentially amounts to your first book-length study. See the writers on your lists as writing mentors of a different sort. When you come across a passage in your reading that seems particularly well-written, an introduction that structures a text cleanly and provocatively, or a sentence that clarifies a difficult point successfully, do your best to think about why it works, and consider marking it for later reference.

4. Develop daily work habits that you can put to use as an ABD. As discussed above, the exams mark the dividing line between periods of highly structured time and largely unstructured time. Once course-work ends, successful scholars demonstrate an unusual ability to impose structure where, practically speaking, almost no structure exists. Ask yourself, before launching into the reading phase of your examinations, what kind of schedule will be most conducive to your own work habits and personal needs. At what time of the day are you likely to be most productive? At what time of day are you basically useless? When do you prefer to exercise? Although I've already detailed the importance of time management elsewhere in this book (see chapter 3), I simply want to stress again here how important it is to establish regular work habits as soon as you step out of the classroom for the last time as a student.

Conclusion

The significant professional and lifestyle changes that occur during the examination stage of the Ph.D. process can cause confusion and stress for many students. While you should expect to do some real soul-searching when course-work ends, you should also approach your exams with a positive attitude and take the necessary steps for avoiding unnecessary delays and headaches. As always, your willingness to understand the purposes behind the practice, and your ability to stay focused on them, will result in a largely positive experience.

Chapter 8

The Dissertation

The dissertation is at once the culmination of various skills learned from day one in graduate school and a completely new monster altogether. Course-work has taught you how to conduct research, to read critically, and to identify common methodological approaches to the materials in your field. Seminar papers have taught you how to write chapter-length analyses of those materials. Examinations have served to expand and focus your knowledge of the specialized area. And hopefully, you will have developed a system for managing your time and research materials that will help you to be efficient and organized. In a very real sense, then, you should possess by the time you are an ABD all of the tools you'll need to write a strong dissertation. However, you've never had to manage quite this amount of material. You've never had to focus on a topic this deeply. Very likely, you've never had to write 300 pages on a single subject.

This chapter argues that the key to writing a successful dissertation lies in the ability of an author to see and approach the final product as the sum of its various parts. You quickly will become overwhelmed, for instance, if you allow yourself to think about writing a three-hundred-page document; instead, think about writing six thirty-page chapters, and you won't feel as anxious since you've written approximately six seminar papers each year that you've been in graduate school. Now add your preface and introduction, a conclusion, and include your bibliography, and you've accounted for your 300 pages. This is a bit oversimplified, but it should give you the basic idea.

In what follows, we'll elaborate on this particular strategic approach to dissertation writing, and we'll cover, in a relatively concise space, dissertation basics from A to Z, including:

- Defining the dissertation
- Selecting a strong topic
- Completing the prospectus

- Filing and organization
- Writing/reading on a daily basis
- Avoiding common dissertation mistakes
- Managing committee members
- Preparing for the defense
- Revising for publication as a book

A good dissertation is perhaps the most important indicator of your potential as a scholar, and to be honest, no document will have a greater impact on how you will be perceived and identified by your colleagues for at least the next decade of your career. While it is crucial that we acknowledge the document's importance, we should also work to keep things in perspective: my main goal here, therefore, is to bring the dissertation back down to earth for you—to demystify it enough to make it manageable and even enjoyable.

What is a Dissertation?

A dissertation is generally defined as a lengthy written treatise on a single subject. In academe, though, the term "dissertation" carries with it all sorts of additional baggage, some of which we should unpack here. The most common adjective associated with the academic dissertation is "original"—as in "an original contribution to the field"—which is highly problematic, as we have already discussed in relation to the seminar paper (see pp. 90–93). While it's true that there's not much point writing something that has already been written, what most dissertators do is explore important topics from new angles or in relation to new materials rather than invent new topics altogether. Perhaps the most useful way to think about the dissertation is in relation to the word's Latin root verb "dissertare" or "dissertate," which means to discuss or debate. Original uses of the noun "dissertation," now obsolete, referred simply to any discussion or debate. The etymology is useful because the roots of the term emphasize that a dissertator is a person who *participates* in a conversation rather than a person who *creates* a specific type of document. The first point worth stressing, then, is that it will be far more productive for you to consider how you might advance a provocative and important conversation than to begin racking your brain for a completely *new* subject of conversation.

Another basic issue of some importance has to do with the differences between a dissertation and a published, single-author book. After all, a monograph (a book written on a single subject) is a lengthy treatise also. Further, any Ph.D. pursuing a serious research career will benefit immensely from being able to turn her dissertation into a book within a few years of obtaining the degree. So should a graduate student think of her dissertation and her first book as one and the same? Dissertation directors approach this issue from different perspectives, of course, but probably too many faculty members continue to regard dissertations and books as completely different species. (On the typical differences between books and dissertations, see pp. 182–87). To be fair, such individuals believe that this more moderate approach serves the best interests of their students, who need not stress out about such things as books before they even earn their degrees. So the argument goes.

I would recommend that you *think* of your dissertation as a book and that you write it in the form and style of a published scholarly monograph. The reality is that your dissertation may—and probably will—wind up looking very different in the end from most published monographs, but such a fact should not be used to rationalize the decision to write a document that will be of little use to you once you are on the tenure track. Besides, one can easily argue that approaching the dissertation as a book helps to cut down on stress, not increase it. For one, students may have a hard time motivating themselves to write a treatise that will be seen by few people other than their advisors. Publishing a book is a more ambitious and inspiring goal, and a student's desire to achieve it can do much to propel her forward. If the goal is merely to write the dissertation in the service of obtaining the degree, then the treatise has already been relegated to a secondary function or, at least, has been defined as a means to an end. You'd be smarter to regard your dissertation as the most important part of your graduate training. Second, Ph.D. students know far better what a book looks like than what a dissertation looks like. In fact, you may find it significantly easier to emulate a document with which you have become extremely familiar than to write a document about which you have only a vague idea. Finally, approaching the dissertation as a book cuts down on the far more significant stress that one may face as an assistant professor upon realizing that one's dissertation will have to be all but completely rewritten. The simple fact is that whereas you have much to lose by writing a dissertation that winds up looking nothing like a book, you have absolutely nothing to lose—and, potentially, quite a bit to gain—by trying to write a book that, at worst, may wind up looking like a typical dissertation.

When you finally achieve that coveted ABD status, then, be positive and ambitious about the research challenges that lie ahead. Imagine what that first book will look like fresh off the press, and capitalize on the energy that such enthusiasm can yield.

CHOOSING AN APPROPRIATE DISSERTATION TOPIC

From the very start of this book, you've been advised multiple times to seize such opportunities as seminar paper assignments and exams in order to prepare for, and even begin drafting, your dissertation. Such advice will only be useful, of course, once you've decided on a dissertation topic, but how does one go about such a process? How does one select a topic that will occupy most of one's time and attention for at least two years and probably many more? And how does one know that the topic is an appropriate one, let alone one that may eventually be publishable in book form? Without question, if such questions were easy to answer, the average national time to completion for Ph.D.s would be much shorter than it is, and attrition rates would be lower than they are. The simple fact is that such questions are not easy to answer, and struggling with them can cause students (and their advisors) considerable stress and anxiety if not dealt with properly.

Do Not *Over*specialize

Hyper-specialization in the humanities may have reached a peak in the late 1980s and early '90s. One could argue that the most marketable dissertations at the current time—while still highly specialized tomes in the eyes of most outsiders—cover a wider range of material than the average dissertation. While cultural history Ph.D.s might still reasonably choose to explore a particular microhistory, or an art history Ph.D. might choose to write on a single architect, it makes more and more sense for Ph.D.s (and especially those working on foreign languages) to select topics that cover a wider period of time and a larger group of materials. The first reason for this is the corporate university's assault on tenure and the increasingly poor ratio of tenured and tenure-track professors to students. Whereas many departments in the past were able to justify separate specialists on Spenser, Shakespeare, Donne, and Milton, for instance, today departments with fewer faculty lines are more likely to hire one or two assistant professors who can teach all four English Renaissance writers (which has always been the case at smaller schools). A dissertation, therefore, that demonstrates an ability to cover the 100-year period between about 1570 and 1670 will sell better than one focused merely on a decade or so. Second, massive cuts to funding for academic presses have diminished the ability of editors to publish books that are unlikely to sell at least some copies (see pp. 216–17). As the editor of one prestigious press told me when I inquired about his willingness to publish my first book (which covered 150 years and included chapters on both Shakespeare and Milton), "We regret that we are unable to publish your book. Clearly yours is an important work, but we are not able to consider such specialized topics at this time." Now maybe he was lying in order to be nice but, without question, this changing publishing market, in addition to the need for new assistant professors to turn their dissertations into books, obviously has had a major impact on the dissertations that Ph.D.s choose to write.

An accidental benefit of such unfortunate developments may be the greater marketability of many dissertations upon their authors' entries onto the job market. Whereas in the past, an overspecialized dissertation sent the message to liberal arts college search committees that a candidate would not be able to meet the needs of the institution, an overly general one may have turned off research universities. One's dissertation topic, in other words, largely dictated the sorts of jobs one was qualified to earn. A candidate on the job market today, however, boasting a dissertation that covers multiple subjects and a wider period of time, may succeed in attracting the attention of both liberal arts colleges *and* research universities. Without question, one should seek to impress twenty-first-century job search committees with a sense of just how much one can cover—as both a teacher and a scholar.

To push a specific example a bit further, consider the following scenario: at a crucial point in time, a Ph.D. in English renaissance literature faces a dilemma about what to write her dissertation on. Her top choices include a study of John Donne's satires or a study of changing perceptions of revenge during the sixteenth and seventeenth centuries. Consider that the latter project would allow the author to cover multiple authors and multiple genres in addition to a wider period of time. Whereas the first choice would allow the candidate to market herself as a poetry specialist, a Donne specialist, or perhaps a specialist on Jacobean literature, the second choice allows her to apply for jobs in Renaissance literature widely construed; historicism; sixteenth-century literature or seventeenth-century literature; some combination of early Tudor, Elizabeth, Jacobean, Caroline, Interregnum, and Restoration literature; in poetry or prose or drama, or all three; in Spenser or Shakespeare or Donne or Milton, or all four. Why would any college or university search committee choose a candidate who can teach and study one poet?

What Love's Got to Do With It

You should be extremely passionate about your topic. Assuming that you eventually move to publish your dissertation as a book, you can expect that the project will be the center of your attention for at least the next 5 years of your life. When the time comes finally to choose a topic, you no doubt will have to think pragmatically about those projects that are most likely to assist you in pursuing your career goals, but such considerations should always be balanced by a more visceral sense of what is likely to continue inspiring you over the long-haul. Such facts suggest the additional advantages of writing on a more wide-ranging dissertation topic—that is, the relative ability to shift focus occasionally or at least from chapter to chapter.

One of my dissertation advisors told me an interesting story on the day of my defense. Having just defended his own dissertation on early modern Italian crime, he returned triumphantly to his little, one-bedroom apartment only to feel himself becoming sick to his stomach. For days he was nauseous and didn't feel like leaving the apartment. But a strange thing happened when his family finally came to pick him up; upon leaving the apartment, he felt better almost instantaneously. When he returned to pick up his belongings from the apartment a few days later, immediate feelings of queasiness confirmed his suspicion: the apartment in which he had written his dissertation—which he would now and forever associate with painful memories of slogging through page after page after page—was making him sick. He moved out.

Every successful Ph.D. probably can relate to such a story (whether it's true or not), but many would also feel compelled to testify that your dissertation need not make you ill. In fact, many ex-dissertators, including myself, would tell you that the dissertation can be the single most enjoyable part of the Ph.D. process. The best dissertations are written on topics about which their authors feel completely obsessed.

Forecast the Future

If you begin your dissertation in 2006, there's not much sense in writing on the hottest book topics from 2006. Remember that even the most efficient dissertator will not be able to publish the project in fewer than about four years time from its conception. The hottest topics from 2006 will be old hat in 2010. Further, you can bet that because most of your competition won't hear this advice and will do what dissertators have always done—write on the hot topics of the day rather than think critically about what is likely to be hot in half a decade—there is very likely to be a glut of dissertations on two or three particular topics while you are on the market. Since your job prospects will hinge largely on your ability to separate yourself from the crowd, you'll want to avoid contributing to such a glut by asking yourself which next step might be worth the taking for a scholar in your particular field.

Your ability to estimate what might be hot in a few years will depend largely on your ability to analyze critically the current trends in your field. Most of the influential work of one period contains hints about what is likely to be influential in the next period. Oftentimes one can discern based on contemporary historical events those topics that will likely be of interest over the coming years. For example, authors sharp enough in the early '90s to guess that millennialism would be a hot topic by the time their dissertations were complete should be applauded. From our current perspective, we can safely assume that the events of September 11, 2001 and the recent U.S. war in Iraq will influence the academic publishing industry, in all disciplines, for at least the next 10 to 15 years.

At the same time that we emphasize the undeniable importance of historical contexts in determining what's marketable and what's not, we should also qualify our statement somewhat by recognizing the simple point that high quality work will always sell better than trendy work. While any good dissertation will speak in one way or another to the particular concerns of the day, you should never write one simply in order to "cash in" on what looks like an opportunity. Doing so would compromise the more important need of the dissertator to write on a topic about which she feels passionate, and it would likely result in a fairly mediocre and transparent final product. In choosing your topic, you'll want to balance almost evenly the three considerations we've discussed in this brief section. Each time you think you've figured it out, ask yourself the following question: will this dissertation topic convey my enthusiasm for what I do, my ability to think intelligently about more than one thing, and my sense of what's important to my colleagues and students? If the answer is "yes," you've got a dissertation topic.

WRITING THE PROSPECTUS

Chapter 2 (pp. 48–49) outlines the potential usefulness of the dissertation prospectus and emphasizes the point that a prospectus should not be approached as a shorter version of the dissertation. Too many Ph.D. students waste an inexcusable amount of time at the prospectus stage trying to solve all of the problems of the actual dissertation. Keep in mind that your dissertation will afford you not only about

2 years of time to solve such problems but also about 300 pages of space for doing so. The prospectus is an approximately fifteen-page document that merely *proposes* what the eventual project will attempt to accomplish. It should set out in clear terms the subject of the dissertation, the manner in which the subject contributes to the author's discipline and area of specialization, and the methodology by which the aims of the author will be advanced. Different universities, disciplines, and individual advisors will employ different practices for guiding Ph.D. prospectuses, but most will expect a prospectus to include the following parts:

The Argument/Thesis Section: While it's unlikely that you'll be able to articulate accurately at the prospectus stage what the eventual conclusion of your dissertation will be, you must seek to offer, at the very least, an informed hypothesis. While most of this two- or three-page section simply announces the subject of the dissertation, your advisors will expect a rather explicit claim (or series of claims) suggesting, first, how you are approaching the subject and second, how your approach responds to and advances previous scholarship on the subject. The first two paragraphs of the dissertation prospectus in the appendix (pp. 300–01) succeed admirably in conveying both points.

The Methodology Section: Dissertators should do their best to explain exactly *how* they plan to approach their materials and fulfill their goals. If you happen to be approaching a particular problem from a feminist perspective, for example, you should say so, since such a perspective is likely to influence how you structure the project as a whole and how you write the document. If answering the questions you have about medieval anchorholds requires that you engage the work of architecture and space theorists, then explain this point to your audience. See the appendix (pp. 301–02).

Work Plan or Calendar: Explain in your prospectus your goals for approaching and completing the dissertation. Some individuals choose to provide an actual semester-by-semester calendar like the one on page 179 of this chapter, while others explain in prose their plans for writing the dissertation, as follows:

> The time period covered by my dissertation is rather large and requires that I become familiar with scholarship and critical discourses that often seem wholly distinctive from one another. Still, I believe that the specificity of my topic will allow me to complete this thesis within two years. The absence of many primary documents in our library's holdings will require that I do several weeks of research at both the Huntington and

> the British libraries, but this research will not entail any major scheduling problems or delays. By considering the introduction and afterword together as the equivalent of one chapter, the dissertation can be figured as seven separate chapters. Two years would allow me to spend nearly a semester on each one. The projected length of the thesis is approximately 300 pages, including prefatory materials and the bibliography.

I discuss in more detail below the importance of long-term planning for dissertators (see pp. 178–79). For now, understand that the prospectus stage is the appropriate time for you to work out such a plan.

Chapter Summary: Chapter summaries should be extremely brief. The point is to suggest which material each chapter will cover, to convey how the chapter fits into the larger project—how it follows the previous one and sets up the next—and to hypothesize about what you expect the chapter to reveal. The prospectus in the appendix effectively demonstrates how approximately two paragraphs constitute more than an adequate amount of space for such purposes (see pp. 302–03).

Bibliography: In some cases, students are not required to submit a complete bibliography of works to be consulted for the dissertation, but there undoubtedly should be one. I recommend that you work up and submit a bibliography regardless of whether or not it's required. Since in order to write the prospectus, you will have researched thoroughly the relevant literature out there on the subject, it won't be difficult for you to build lists of the primary and secondary works likely to inform your dissertation. Such lists will indicate to your committee members how much research you have done, and they will serve a number of practical purposes for you later on. For one, they suggest with which texts you might begin your reading for each separate part of the project. Each time I began a new chapter of my own dissertation, I started by consulting my prospectus bibliography and then collecting from the library all of the works relevant to the particular chapter. Second, by marking next to each entry the call number or location of the articles or books you include, you'll save yourself a lot of time later on in trying to track down (and then re-track down) all of these materials. Obviously, you'll continue to expand the lists that you generate for the prospectus as you move more and more deeply into your dissertation, but few documents will prove more useful to you throughout this long process than a well-constructed prospectus bibliography.

After your prospectus is approved, take it down to a copy store and pay a few dollars to bind two or more copies. Since you'll be using the document for a variety of purposes and probably writing all over it for 2 or more years, you'll benefit from owning a sturdy and portable copy. You may also wish to give a bound copy to each of your advisors so that they can access it more easily when the need arises.

FILING AND ORGANIZATION

If you've read the earlier parts of this book, especially chapter 3, you probably have a pretty good sense by now of how you might organize your materials. Nonetheless, because the dissertation is such a large undertaking, and because I believe organization to be one of the most important—if not the most important—key to writing a strong one, I would like to offer here a few additional comments about organizational issues unique to the dissertation.

You'll need at least a three- or four-drawer filing cabinet for your dissertation materials, which will accumulate more rapidly than you can imagine. Designate and label appropriately two files for each of the following categories: "Prospectus," "Prefatory Materials," "Introduction," "Chapter One," "Chapter Two," and so on, until you have constructed two files for each section of your dissertation. Whereas one of the "Chapter Three" files can be used to store different draft versions of the third chapter, the second "Chapter Three" file can be used to store any articles, notes, or photocopied book chapters you've used as part of your research for the chapter. Make it a point to save one copy of each significant draft you print out for each chapter; since you'll constantly be cutting, altering, and extending chapter material, you'll want to keep track of the various changes you make, and you'll want to maintain access to deleted ideas, even single sentences, that you are unlikely to save in separate computer files. About the secondary literature files, my advice is that you copy everything you consult and certainly everything you quote in your dissertation. You'll need to return to these files repeatedly, not only when you wish to expand a passage or introduce a new one but also when you face more mechanical tasks such as proofreading published versions of the dissertation.

It may also be advisable for you to establish separate files for central subtopics or themes in your dissertation. For example, since my own dissertation was significantly informed by scholarship on early modern bodies, and since this scholarship was applicable across all six of my chapters and the introduction, I set up a separate file labeled "Body Criticism." I also established a file for "*OED* Entries" and other reference materials. Eventually I added files for professional issues pertaining to post-composition matters, such as "Defense Preparation

Materials" and "Potential Publishers." As always, you should tailor your filing system to your own individual needs, but don't assume it will be easy to survive writing the dissertation without some kind of system.

THE DAILY LIFE OF THE DISSERTATOR

What most new ABDs really want to know about the dissertation, of course, is how to approach the actual process of writing it. For the first time in your graduate career, you will begin to think and feel like a real writer, and how you approach your writing on a daily basis will affect everything from your psychological health to your time to completion. While the subject is worthy of book-length attention, I offer in this brief space what I regard as the five commandments of dissertation writing:

Slow and Steady Wins the Race: As we discussed at the beginning of this chapter, a dissertation can be a terrifying thing to confront if approached in its entirety. In small parts, however, the dissertation can be conquered quite painlessly. The most important step you will take in order to stay focused on one thing at a time involves arranging your schedule in a manner reflective of this basic principle. Dissertators who sit down at their computers every few months and try to dash off 40-page chapters usually wind up rewriting those chapters. They also mature very little as professional writers because they refuse to spend the time it takes to focus on the minute details of involved research. Your first goal, therefore, should be to write a small part of your dissertation every single day. This is harder at first. Obviously you cannot write if you have done *none* of the reading for a chapter, but in fact you did lots for the prospectus; if nothing else, you can organize your notes from that document, lay out the pattern of reading best suited to building the chapter, and begin drafting an introduction.

To this day, I continue to practice the daily writing rituals that I developed years ago as a new ABD. For me, progress is best measured by actual material output, and so I was determined, as a dissertator, to write a minimum of two pages every single day. Since I scheduled to teach in the afternoons or evenings, I was able to sit down at my computer each morning with a clear and focused goal in mind. On some mornings I would write my two pages in less than an hour, and then I might choose to shift my attention to other matters. On some other mornings, I would feel so energized by my progress that I would continue to write—sometimes as many as ten pages. And yet on other days, it would take me all day (and part of the night) to get out two pages. But I always stuck to my rule. Now think about this: 2 pages is only about 500 words. But writing 2 pages a day over a year's time will result in a roughly 700-page manuscript. Now even if you estimate conservatively that you will wind up cutting about half of the material you generate, you still will have produced a document longer than 300 pages. Other writers think less in terms of material output than in terms of time put in. For instance, many people make it a point to write for a minimum of two hours a day. During this time, they may write two pages or they may write five, but they are always writing.

Besides sanity, productivity, and the establishment of routine, a major benefit of such daily practices is the improvement of your writing—stylistically, rhetorically, and grammatically. By deciding to focus on two pages a day, or perhaps three pages for two hours, you allow yourself to study the various building blocks of the larger structure you're creating. If you walk into a museum and, from a distance, stare at a massive painting for an entire afternoon, you'll no doubt learn a lot about the painting. If in an alternative scenario, you focus for two hours, for five days in a row, on a very small section of the canvas, you'll learn a lot more about it—about the manner in which the paint has been mixed and layered, about the variety of colors used to create the image, about the way that light and shade are used to create depth, and so on. The first observer of the painting spends about the same amount of time studying the image and yet possesses an inferior understanding of how it was created and probably would have a very hard time describing it intelligently to persons knowledgeable about art. The second observer knows the object inside out and could probably describe it in an impressive amount of detail. The point is that once you understand how to bring the small parts of your own project together in order to produce a desired effect, you'll be able to replicate that process quickly and purposefully as you continue to move forward.

I do understand, of course, that many people will find it more desirable to research and read everything up front and turn to their writing only after they have begun to feel more comfortable discussing the dissertation material. Such persons may feel that there's little to be gained from generating much prose at an extremely early stage, which may be true for them. Only you can decide what works best for you as a writer. What I would try to take away from the two-a-day models, however, is the simple point that you should be working on your dissertation—in one way or another—every single day.

Stick to Your Long-term Goals: A long-term plan for completing the dissertation will help you to measure how well your daily writing practices are working. In chapter 2, we discussed the reasonable goal of trying to finish your dissertation in 2 years time. Assuming that your dissertation will break down into about seven chapters (counting the introduction as one), it would be wise to spend approximately three months on each one. Now time in the summer is different from time in the fall semester (obviously, there's more of it), and so you should schedule your research as realistically as possible. I recommend that you aim to write a single chapter each semester that you are teaching and two chapters in each summer, so that your two-year schedule would look something like the one proposed in figure 8.1.

While such a schedule would be realistic even for an ABD truly *beginning* her dissertation after comps, it's decidedly more realistic for those individuals who are strategic enough to use their final few courses to draft a chapter or two. Further, if you stick to your plan of writing one chapter each semester, you'll be able to use the winter holiday break as a productive time for launching into the next one. Finally, if your university happens to grant you a semester reprieve from teaching at the dissertation stage, as many do, you'll be able to get even further ahead of schedule.

In any case, hold yourself to the deadlines you establish. If you feel that you need help doing so, bring your long-term schedule to your dissertation advisor, explain that you would prefer to be held to it, and ask that she demand work from you every three months or so. If your primary advisor is a softy, seek the help you need from other advisors or writing group peers. By combining short-term (i.e., daily) and long-term plans for writing your dissertation, and following them religiously, you should be more than able to finish the project in about two years time.

Know When To "Stop" Reading

My advisees always tell me that what hinders their progress most is the overwhelming feeling that there's always more out there to read.

Spring year five:	Defend prospectus and write introduction
Summer year five:	Chapters one and two
Fall year six:	Chapter three
Spring year six:	Chapter four
Summer year six:	Chapters five and six
Fall year seven:	Revisions and defense (prior to job market interviews)

Figure 8.1 Sample 2-year dissertation plan

I should say that it's not just a feeling. There *is* always more out there to read. But by dichotomizing reading and writing as separate activities, and especially by assuming that one needs to read everything *before* beginning to write, dissertators unnecessarily slow their progress. The only time when one really needs to read (almost) everything "up front" is prior to the prospectus/introduction stage. Once you actually begin writing, though, you must make it a point to keep writing *while you're reading for the next part.* Figure 8.2 suggests what a realistic weekly schedule might look like for a writer close to finishing her first chapter.

Notice how conducive a two-page-a-day system is to maintaining such a weekly schedule. A major incentive for finishing your two pages (or your two hours) by, say, 10:00 AM, is the amount of time you will keep free for reading and revising. Another advantage of such a schedule is that the materials you read will be fresh in your mind while writing about them. If, on the other hand, you were to read during the entire month of July and then try to write in August and September, your memory of what you've read would likely be compromised. Finally, you should remember that it's never too late to read new materials for a particular chapter. Your advisor will regularly suggest upon reading your drafts that you add a certain article or mention a certain book, and new publications by other writers will force you to update chapters long after you finish your first drafts. Read thoroughly, but don't allow your fears about having to read everything hinder your progress.

Consider Writing Your Introduction Last

While in some cases, such a move will be counterproductive—such as where a base needs to be established before anything else can make sense—many dissertation writers will benefit from waiting until the very end to write their introductions. The simple fact is that you can't

Mon:	Write two pages of chapter one; finish final reading for chapter one.
Tues:	Write two pages of chapter one; begin reading for chapter two.
Wed:	Write two pages of chapter one; continue reading for chapter two.
Thurs:	Write two pages of chapter one; continue reading for chapter two.
Fri:	Finish chapter one draft; continue reading for chapter two.
Sat:	Read for chapter two.
Sun:	Read for chapter two.
Mon:	Begin introduction for chapter two; continue reading for chapter two; begin revising chapter one.

Figure 8.2 Hypothetical weekly dissertation schedule

possibly know what your conclusions (i.e. your major claims) will be until you have completed your research. Many dissertators get bogged down trying to articulate answers before they've even begun to ask the right questions. In order to avoid the common problem of the never-ending introduction, consider diving right in to that section of your dissertation about which you feel most comfortable. Once you experience the momentum that comes from completing a first chapter, and once you begin to see the sorts of answers your inquiries are likely to turn up, you'll find it much easier to introduce your dissertation.

Approval Is Not Required

Often ABDs who turn in a draft of their first chapter will stop writing until they hear back from their advisor. This is an unproductive personal policy for several reasons: first of all, a dissertation chapter is simply part of a larger project; it's important, therefore, to channel the momentum you've established in order to complete one chapter into the next chapter. Doing so will enhance your sense of the connections between the numerous parts. Second, your advisor may take weeks to read your draft, valuable time that you cannot afford to waste. Most important, though, waiting for an advisor's approval implies that you view a draft as a finished document. You can be sure that you always will have revisions to perform once your advisor reads over your drafts. Sometimes these revisions will be trivial and sometimes they will be major. So if you're going to have to revise one way or the other, why halt your progress?

The problem is compounded when students insist on showing every chapter draft to every one of their advisors. While some departments and/or major advisors require that all committee members review all chapters as they are produced, most do not, and you would be wise to work alone with your major advisor until she has approved each chapter for review by your other committee members. Most secondary committee members prefer such a system because it saves them a considerable amount of time from reading and commenting on material that eventually will be trashed. Obviously if one of your secondary advisors happens to be an expert on a particular issue engaged by one of your chapters, then you should seek that person's advice and feedback earlier. Otherwise, allow your major advisor to guide you through the project, and go to the other advisors only when the end is near.

Nighttime Is the Right Time

A dissertation doesn't consist only of chapters. Like a published book, it also includes a title page, acknowledgments, a table of contents, a bibliography, and so on. While some really well organized individuals will budget their time so effectively that there will be several weeks left over for them to work on such sections, most of us will be forced to cram them in while trying to complete more complex sections of the dissertation. Try not to waste too much time during the day—during your most productive writing time, that is—working on such things. Near the end of my own dissertation career, I began working on such materials at night while watching the news or sitting outside on the porch with a drink. Unlike searching for the most persuasive formulation for conveying an important idea, say, writing a table-of-contents page or revising a bibliography for mechanical consistency are not activities that require a great deal of brain power. They might require a good deal of your time, though, if you fail to plan appropriately for them. About three months prior to your submission deadline, therefore, you should make a checklist of all of the documents required of you from your committee, your university, and yourself, and begin knocking them off whenever you have a little free time.

Avoiding Common Dissertation Mistakes

At the beginning of this chapter, we discussed that every dissertator's goal should be to write a book—not a dissertation. So what's the difference between the two, you might ask? Dissertations tend to be characterized by one or more problematic features that prevent them from being taken seriously by editors. A basic awareness of such potential pitfalls and a serious effort on your part to avoid them can result in an infinitely more publishable dissertation.

Advance the Conversation; Don't Prolong It

The most common, and commonly acknowledged, problem with dissertations is that often they belabor the "situating" move (see pp. 105–07) to the point of absurdity. For readers, getting through such sections of dissertations can be extremely tedious and frustrating. We might explain the frequency with which this problem occurs in one of two particular ways: either the author lacks the ability to discern what her readers will need to know in order to follow and find persuasive her argument, or she lacks the confidence necessary to say something on her own, as opposed to describing what's already been said. The first problem is less excusable since Ph.D. candidates should possess by the time they are ABD an acute understanding of what their readers know and what they will need to be taught. An audience of Michelangelo scholars, for example, won't need to be reminded that the statue of David was originally unveiled in the Piazza della Signoria and not the Galleria, yet, many (stereotypical) dissertators will use a considerable amount of space on such simple observations. To be fair, the example is extreme, but I'm often surprised by how much *obvious information* needs to be cut from typical dissertations. The second cause—insufficient confidence—is more complex, as the next section recognizes.

Let Your Voice Be Heard

As discussed in chapter 5 (pp. 108–14) an author's anxieties about his own credibility can lead her to subordinate her voice to the point where it cannot be heard at all. Sometimes even the most original and provocative arguments are difficult to locate because of authorial tendencies to overcompensate or be overly polite. It certainly does not help that so many writers harbor misconceptions about the appropriateness or inappropriateness of using the first person singular voice. Remember that the first person is only problematic when used in a way that weakens an argument by diminishing the illusion of objectivity: that is, when an author begins a sentence with "I believe" or "In my opinion." Using the first person voice to describe what you're arguing and how you're setting out to do it, however, is not only acceptable—it's also advisable. By embracing grammatical formulations that convey your views explicitly, you can train yourself to keep your voice central and others' voices peripheral or supportive. Consider the following example of an opening paragraph from a published article by Valerie Traub, entitled "The (In)Significance of 'Lesbian' Desire in Early Modern England":

> The "lesbian desire" of my title is a deliberate come-on. If this is the last you hear of it, it is because, enticing as it may sound, it doesn't exist. Not, at least, as such. For the conceptual framework within which was articulated an early modern discourse of female desire is radically different from that which governs our own modes of perception and experience. . . . [H]ow is the . . . recent discursive invention, the lesbian, to be related to sexual systems of four hundred years ago? The following discussion attempts to begin to answer that question. . . . My intent is to keep alive our historical difference from early modern women and at the same time to show how historically distant representations of female desire can be correlated . . . to modern systems of intelligibility and political efficacy. This essay is at once an act of historical recovery and a meditation on the difficulties inhering in such an act.[1]

Notice how strong and unmistakable is Traub's authorial presence here. No reader could put down the article after reading its introduction without a fairly clear sense of what Traub hopes to accomplish. Such clarity is a basic element of most well written articles. As a dissertator, your goal should be to convey the same sense of clarity and conviction as a published author, and to recognize that your failure to do so will hinder your dissertation from becoming a book sooner rather than later.

If the problems plaguing you are less grammatical or rhetorical than they are psychological, then it's time to sit back, take a deep breath, and recognize just how credible a voice you happen to own. When you're feeling particularly anxious, try to remember a few simple points: you have completed several, if not many, challenging classes focused on the general field you're researching; you have earned enough admiration from multiple experts in your discipline to convince them to serve on your committee; you have passed challenging examinations in the given subject area; you have received approval of a detailed dissertation prospectus from several discriminating readers; and, most important, because your work will be submitted to a committee, you can be relatively certain that no risks you ever take as a maturing writer are likely to go unchecked by experts who are seriously invested in your future success. No one can deny that it takes courage to be a writer—courage, first of all, to feel so passionately about a subject that you're willing to dedicate years of your life to it; second, courage to join a conversation so complex that it would simply scare off or silence most persons; and finally, courage to put your ideas into circulation where other intellectuals can test and even challenge them. Trust in those around you and trust enough in yourself to go out on that limb. The thrill of standing quickly will overshadow the fear of falling.

Argue, Don't Catalogue

What I would call the "taxonomic approach" to dissertation writing usually involves the announcement of a rarely discussed subject in the introduction, followed by chapters that categorize or simply organize different manifestations or parts of that subject. For example, a scholar might decide to write a dissertation focused on portrayals of New York City in modern film. A taxonomic approach to such a dissertation would result in something like the following structure: an introduction explaining that too few critics have looked at cinematic depictions of the city (not really true, of course); five or six chapters arranged in categories such as "New York City as Slum," "New York City as "Melting Pot," "New York City as Paradise," and so on, all of which merely describe the different "types" on which the author has chosen to focus. Such a dissertation might help to "point out" trends in film making, but it would probably do very little to advance our understanding of how film works or why filmmakers choose to depict the city in certain ways.

In the humanities, as we said earlier, dissertations should be discursive or argumentative, not merely descriptive. An improved dissertation might argue something like the following: "1970s Hollywood directors have deliberately rejected the idea of New York as a city containing multitudes, opting instead to portray it from the stereotyped perspectives of a single ethnic group." Rather than *announcing* a subject, the author introduces the dissertation by offering us a clear claim (X = Y). Later, when she will go on to explain the basis for her claim (i.e., the "because" element, or the reason why Hollywood directors have constructed only a certain type of city), her argument will have been completely established, and she will be ready to begin backing it up by turning to chapters with titles like "African American NYC," "Italian American NYC," "Irish American NYC," and so on. In this case, our writer is saying something about film that might help us to understand how Hollywood filmmakers operate or how certain types of films serve the economic and political aims of the Hollywood system. In short, in this case, the author is actually saying something, though she's not quite out of danger yet.

Write Chapters That Advance the Argument

Whereas in the case of the taxonomic dissertation posited above, the chapter titles emphasize an extreme variety of directoral possibilities, attempting to point out all of the different ways that a director *could* depict New York City, the chapters of the discursive dissertation are more selectively chosen to advance the argument established early on in the introduction. Still, even the second dissertation will be rather weak if its chapters only succeed in *repeating* the argument articulated in the introduction. For example, one can easily anticipate a first chapter that begins like this: "In the Introduction, we saw that Hollywood directors have tended to portray New York City from several reductive, even stereotyped ethnic perspectives; in this chapter, I show how many directors film New York City from the perspective of disenfranchised African Americans." Now even if the author succeeds in accomplishing the goal, her readers will learn very little from the chapter that they haven't already learned from her introduction. Whereas the general introduction presents the argument in general terms, the chapter introduction presents the same argument from a more specific angle. If you're not seeing the problem here, you might consider that whereas the general introduction presents us with a panoramic view of Manhattan, the first chapter zooms and crops to the point that we are only able to see one or two neighborhoods of the city. The picture is exactly the same, though.

Unlike authors of most published monographs, dissertators often repeat themselves in each chapter of their project. Be aware of this common mistake, and seek in every chapter to keep advancing your argument. Readers of academic treatises must always feel as though they are continuing to learn something about a subject or they simply will stop reading. In this regard, the basis used to support the central claim is perhaps the master key to ensuring that each chapter advances the project's argument. For example, let's imagine that the author of our dissertation on New York City films contextualized her claim in the following manner:

> Claim (X = Y): *1970s Hollywood directors deliberately rejected the idea of New York as a city containing multitudes, opting instead to portray it from the stereotyped perspectives of a single ethnic group.*
>
> Basis (because): *Such a strategy amounted to an attempt on the part of movie producers to exploit the unique anxieties of certain ethnic and racial communities after the breakdown of segregation in the 1950s and '60s.*

A strong dissertation would now highlight within each chapter first how specific ethnic groups actually responded to the civil rights movement, historicizing the anxieties that the author claims they experienced, and second, how specific films spoke to each community's particular concerns. In such a dissertation, individual chapters are unlikely to repeat one another because each community's concerns are said to be unique; therefore, each chapter will discuss matters in relation to specific local contexts, though together, the chapters will demonstrate the validity of the author's global claim.

Eliminate Jargon and Rhetorical Convolution

I won't insult your intelligence by claiming that academic monographs are always clearly written, but I will insist that most of them are so. While some dissertations are poorly written simply because their authors have not taken the necessary steps toward developing clear and concise arguments, convolution and confusion often are direct results of authorial attempts to construct a voice that sounds "academic." While an author should always be aware of the rhetorical and compositional modes that she sees dominating her specialized field, she should be wary of inventing a voice unnatural to her own style or ways of reading and writing. Despite the nonsensical claims of right-wing critics of academic writing such as Lynne Cheney, it will not always be possible to convey your ideas in terms that your cousin Sally will be able to understand. Academic authors engage highly specialized topics, and their complex arguments demand complex formulations. Nonetheless, as a practical strategy, I recommend that you *think about* Sally while writing your dissertation. How might certain compositional and rhetorical decisions enhance the clarity of your writing? How might certain decisions increase your persuasiveness?

This is not a book about writing, and you may not be in need of such a book. If you are interested in avoiding common mistakes made by dissertators, though, or if your committee members keep imploring you to clarify or explain your ideas better, you will benefit greatly by consulting books on the academic writing process. If the problems you're facing are mainly stylistic, I would recommend that you read Claire Kehrwald Cook's *Line by Line*, written specifically for scholars, and Richard Lanham's *Revising Prose*. If you're making mechanical errors, skim *The Chicago Manual of Style* or the *MLA Style Manual* (not the same as the *MLA Handbook*) from front to back, and mark the relevant pages clearly so you can return to them when necessary. If you need a grammar lesson, Strunk and White's *Elements of Style* is still one of the most concise and affordable guides. Finally, if you believe that the writing problems you're experiencing have more to do with the unique problems of writing a dissertation, consult Patrick Dunleavy's *Authoring a Ph.D.*, which covers everything from discovering a topic to defending and publishing the project; David Sternberg's classic *How to Complete and Survive a Doctoral Dissertation* offers invaluable advice about how to organize dissertation materials and ideas.

By keeping in mind and seeking to avoid the common mistakes made by dissertators, you can bring your dissertation that much closer to publication; you can also eliminate the time-consuming revisions that so many individuals are forced to make while on the tenure track—a time when they should be focused on making more substantive revisions and trying to find a publisher for their work.

MANAGING THE ADVISORY COMMITTEE

If you're lucky, your advisory committee will consist of three or four extremely intelligent people who happen to disagree about a lot of things—since it won't be helpful for you to see things from only a single perspective. At the same time, a considerable amount of stress often is caused by disagreements between committee members. Worse, such disagreements can delay dissertations from being written in a timely matter. In what follows, I offer some advice about how to cut off committee problems before they get out of hand.

The Major Advisor is Boss

The most important thing for every dissertator to remember is that, should any conflicts arise, the primary advisor is responsible for resolving them. As mentioned above, you can avoid many problems and delays by submitting the work in progress to the major advisor alone and then submitting the approved chapters to the rest of your committee members later on. Such a strategy sends a message to secondary advisors that the project has already been approved by the major advisor, which reduces the likelihood that they will quibble over trivial matters or minor differences of opinion. I am not suggesting, of course, that you use the major advisor's authority to avoid having to make what may be productive revisions, cuts, or additions; I am suggesting that you write the dissertation that you and your major advisor have agreed you would write.

Decoding Mixed Messages

There will be times when two committee members will give you contradictory advice. When I was writing my own dissertation, I learned from one of my secondary advisors that a particular claim, which my major advisor had described as the "crowning achievement" of the project, was actually "ill-informed and damaging to [my] overall argument." (This never ends, by the way: after publishing the revised version of my dissertation as a book, two separate reviewers described the same exact fifth chapter as both the "weakest" and the "most excellent" part of my project!) Such conflicts can be difficult for dissertators who obviously would prefer not to seem disrespectful of any one advisor's opinion. Further, it's not always clear which advisor is really correct or whose advice should be heeded. Here's how to handle such a situation: (1) First of all, recognize that the more negative assessment may be valid. As a writer with a deep investment in this work, you will always be more inclined to take seriously the positive comments since doing so means avoiding revisions. Keep in mind, though, that if the more critical comment is indeed valid, revising the passage now may save you from being embarrassed later on—whether by a journal/book editor or, worse, a book reviewer; (2) talk openly to both advisors about the conflict. Simply explain that you've received contradictory advice and that you are sincere about wanting to make the right decision. By tackling the problem in such a way, you'll invite both advisors to review the passage, think more carefully about their assessment of it, and consider a view of it different from their own. Nine times out of ten, an advisor will seek a compromise position. Your dissertation will be better as a result; (3) be willing to surrender when the disputed idea, method, or passage is not crucial to your dissertation. When I first began my dissertation on Renaissance literature, I was determined to frame every chapter with a modern anecdote that would serve as a lens through which to view early modern phenomena. My advisor explained that while she was willing to entertain such a strategy, it didn't make a lot of sense to her and, in fact, it might weaken considerably my historicist approach to the materials under discussion. After some soul-searching, I decided to drop the modern frames and to follow her advice. My reasoning was that I could always add these frames later for the book—when the final decisions would all be mine. After writing the dissertation, however, I came to see that her advice simply was correct: including the frames would have been disastrous. By respecting my advisor's superior sense of how things happen to work in our field, I avoided wasting time and energy on a bad idea; (4) submit irreconcilable differences to the major advisor. If your advisors ever seem unwilling to compromise, you will be in the difficult position of having to ignore the recommendation of one of them. I have two separate thoughts about this fact: the first one is "oh well." No one ever said that the dissertation was going to be easy. As a writer you have to make tough decisions, and you have to be willing to stand up for what you believe to be a good decision. My second thought is "oh no," since sometimes the decisions that you make will offend and alienate committee members. In cases where differences seem irreconcilable and where you're simply not sure what to do, seek the counsel of the major advisor. Her job is to step in and help you to deal with such difficulties, which, I promise, will eventually be resolved.

Problems with the Major Advisor

Most dissertation committee problems can be solved so long as the dissertator's relationship with the major advisor continues to be strong. But what would happen if a conflict were to emerge between you and your major advisor? As a professor and director of graduate studies, who listens regularly to students and their advisors, I can tell you that such conflicts are *extremely* common. I can also tell you that they usually are easily resolvable if both parties are willing to communicate openly about the cause of the conflict. Should a problem ever arise, try first sitting down with your advisor in order to address the matter head-on: "I'm feeling really confused about this issue. On the one hand, I think that this section is the most important part of my dissertation—at least to me. On the other, I have tremendous respect for your advice and can see some of the problems you're pointing out. Is there any compromise to be had here? In other words, isn't there a way I can make this work?" Most advisory committee problems would go away rather immediately if the involved parties would swallow their pride and simply discuss the matter. When such discussions break down, a third party should be consulted. Usually, the best person to speak to is another committee member, though the director of graduate studies can also be consulted. In certain extreme cases, where the relationship between student and advisor has become strained and no immediate solution is apparent, it will be advisable for the student to switch major advisors (or if the problem is with a secondary advisor, to drop her from the committee). Such situations are extremely uncomfortable for everyone involved (and may have consequences when you are on the job market) and should be avoided when possible—but they do arise more often than you'd think.

In the vast majority of situations, however, advisors and their students get on very well together. The minor disagreements that inevitably emerge from time to time during one's dissertation stage will seem like major ordeals to you at the time and laughable trivialities a few weeks later. Simply do your best to communicate your ideas and concerns openly and respectfully, and it's unlikely you'll run into any significant problems.

Marketing the Dissertation

Almost as soon as you begin writing it, you also should begin to "market" your dissertation. The goal should be to create a situation wherein, eventually, any scholar in your field who thinks about the topic of your dissertation also will think immediately of you. What the successful realization of such a goal requires is that you conference, publish, and simply talk with important persons about the various parts of your dissertation. By creating a buzz about the work you're doing, you'll increase the chances that you'll be able to publish it soon and that it will be well received.

Conferences are especially useful events for dissertators. After deciding on a topic, you should pay careful attention to the upcoming conferences in your field and discipline. Any that seem relevant to your work—especially the major national and international ones—should be regarded as opportunities for advertising your work. You should make it a point to present conference-paper versions of every chapter of your dissertation, some prior to graduation. In each of these presentations, you should explain the argument within the context of the larger project so that your audience can gain a better sense of what you're working on and what you're arguing about.

At the larger conferences, you also should make it a point to chat with editors, who often represent their presses at book displays. As I also discuss in the following chapter (see page 211), you should talk to editors about your work to see how they respond. The idea isn't really to pitch the book just yet, but rather to establish relationships with editors and to assess how interested they seem to be in your project. Often they will offer valuable feedback about how such a book should be written or those presses that might make appropriate publishers. Sometimes they will ask you to send them a prospectus or to stay in touch as the project continues to develop (be sure to ask for that person's business card, and keep it). If this should happen, send the editor an e-mail when you get home, thanking her for her time and promising to stay in touch. Later if you decide to ask the editor whether the press would be willing to review your manuscript, you can remind her of her initial interest in the project, which may prevent her from simply dismissing your letter of inquiry.

If you know that persons of some importance to your work will be present at conferences in which you are participating, make it a point to attend their sessions and introduce yourself. Often it may be possible for a common acquaintance, especially an advisor, to introduce you to this person. While still a graduate student, I benefited multiple times from attending conferences where my major advisor was delivering a paper; she was always more than willing to introduce me to people in the field and, on several occasions, I was able to sit down for a drink with these people, whose advice and support were invaluable to me in a variety of ways.

Finally, you should aim to land at least one chapter of your dissertation in a major journal prior to going on the job market, a recommendation I repeat in chapter 10 (see pp. 214–16). Ideally, you should place two articles, but you should be careful not to publish more than about 25 percent of your dissertation, as publishers will balk at reprinting material that's already easily available.

THE DOCTORAL DEFENSE AND BEYOND

Once you submit your dissertation to your committee members, the hardest part is over. You now have three very clearly defined goals: to defend the dissertation successfully, to submit the final version of the document to the graduate school, and to begin the process of revising it for publication as a book.

I've witnessed only one defense in which a Ph.D. candidate failed to defend her dissertation satisfactorily (if it makes you feel any better, the individual was *badly* under-prepared, generally incompetent, and deserved to fail). For most dissertators, though, the word "defense" is not applicable at all. "Celebration" would be a more appropriate term for describing what happens at most of these affairs. The shift toward professionalism in graduate school is partly responsible for the death of the "defense," since most graduate students today write far more complex documents than the dissertations of 50 years ago. When I showed up at my own defense having had published two and a half chapters of the dissertation, I understood that the idea of having to defend the document *as a dissertation* seemed just a little bit weird. Further, most academics tend to agree that sending an unprepared student to her defense constitutes a major gaffe on the part of a dissertation committee and especially the major advisor; they often end up looking far worse than the student when the latter gives a bad performance.

While you should enter your defense with the attitude that you *will* pass, you also should assume that you will be asked some hard questions, and you'll want to be prepared to field them. While the advice I gave you about preparing for the oral defense of your exams (pp. 163–64) is applicable also in the case of the dissertation defense, several steps should be taken to ensure that everything goes well:

1. First of all, make it a point to attend classmates' defenses prior to your own (unfortunately, some universities prohibit non-committee members from attending doctoral defenses). Seeing how different advisors run different defenses will help you to prepare for what you may encounter during your own (and you may even get to see one or more of your own advisors in action).
2. Talk to your major advisor ahead of time about the format of the defense. How will it be run? How should you prepare for it? Will you be expected to say something to the audience prior to the questioning? What does your advisor think is the purpose of a defense?
3. Reread your entire dissertation. While at this point you will likely have huge sections of your dissertation all but memorized, you should enter your defense with a sense of the project as a book or a treatise, not a compilation of individual chapters. Most questions will require that you think about the project as a whole, not in parts.

4. Reread at least the most relevant sections of the major critical works that inform your dissertation, and also the major primary texts that your dissertation analyzes. You'll want the facts to be as fresh in your mind as possible.
5. Rehearse (i.e., memorize) a thirty-second, a two-minute, and a five-minute summary of your dissertation's major argument and how it contributes to previous and ongoing scholarship in your field. These summaries will also be very useful to you during job interviews (see page 268).
6. Finally, think carefully about what you will need to do in order to revise the dissertation for publication as a book. At least one of your questioners is very likely to ask how you plan to revise it. As important, your willingness to recognize the imperfect state of the document, and your plan for addressing its imperfections, will satisfy your critics and perhaps even convince them that more (public) criticism is unnecessary.

Once you've answered all questions to the satisfaction of those persons present at the defense, everyone but your committee members will be asked to leave the room. Stay relaxed knowing that you have done all that you could do. The door will be opened in five to ten minutes—since committees often discuss a bit and then sign a stack of forms, which takes time—you will be informed that you have passed the defense, and the real celebration will begin. Usually privately, your major advisor will explain to you what was said after you departed from the room, and she will tell you those final revisions that you'll need to make before the final version can be submitted to the graduate school. Sometimes the mandatory revisions are minor and sometimes they are more substantive, but, almost always, they can be completed in a relatively short period of time.

Your graduate school will have its own requirements—mainly formal—for how the final version of the dissertation should be formatted and submitted. Most universities enforce rather strict guidelines for ensuring the mechanical consistency and professional quality of all doctoral dissertations. Hi-tech universities have thesis packages with computer templates for putting the dissertation into suitable form. Use them. In addition to at least one copy of the entire project, you will be asked to submit multiple signature forms, a copyright agreement, order forms for binding preferences and extra copies (order a bound copy for each committee member and however many extra copies you would like for yourself and family members), and the UMI abstract.

This 350-word summary of your dissertation—which is extremely painful to write—will be sent to Proquest Information and Learning, where it will be registered with and published by Dissertation Abstracts International. Pick up an instruction booklet from your graduate school early on in your dissertation-writing career. By following the university's guidelines from the beginning, you'll cut down on the work you'll be forced to do later on.

Once you submit the dissertation to the graduate school, the only steps left to take are to order your doctoral cap and gown ($500 to $1,000, by the way) and to register for graduation. The day that you thought would never arrive is here at last. You'll want to begin thinking about those revisions and your first book very soon (see pp. 229–37 on writing the book prospectus), but for now be sure to take a break and think about all that you've accomplished.

CHAPTER 9

ATTENDING CONFERENCES

The thought of volunteering to deliver an in-depth presentation before a couple dozen experts in your field may horrify you, but attending conferences is one of the routine activities of humanities scholars. The actual usefulness of conferences is a more controversial issue. Personally, I am of the viewpoint that well-used conferences can positively transform one's research and even one's career. Nonetheless, in this chapter I will try to show you both sides of the picture. Among the other subjects I will discuss are the following:

- Using oral reports as practice for conferences
- Finding the right conference
- Contacting a conference organizer
- Writing a lively presentation
- Networking
- Conferencing as job market preparation

By and large, you will find that conferences are relatively collegial—even pleasant—affairs and that the anxieties they provoke in most inexperienced scholars are unwarranted. By the time you finish reading this chapter, you should be able to envision what will happen when you attend your first few conferences, which should in turn give you the right amount of confidence to be successful.

WHAT IS A CONFERENCE?

Academic conferences have many different faces, but the one unifying link between them is a very basic goal: to bring together scholars for the purpose of sharing *ongoing* research. From the perspective of the presenter, conferences represent an opportunity to publicize one's work and to receive feedback that may be helpful in the process of revising it. From the perspective of the auditors, conferences represent a type of continuing education, revealing what's hot at the moment in a particular scholarly field and, in the best situations, suggesting likely trends for the foreseeable future.

Conferences are arranged, sponsored, and run by a variety of organizations and individuals. The smallest conferences, usually hosted by individual faculty members in specific college or university departments, may attract only 20 or 30 attendees and may last no more than a day. Such conferences tend to be highly focused on particular topics (e.g., "Minds and Bodies in Medieval Europe"), and they have the potential to foster useful professional relationships by virtue of their ability to draw like-minded people into relatively intimate environments. Universities also host larger, multi-day conferences, which, if conveniently located or focused on a provocative enough topic, can attract hundreds of attendees. These larger conferences feature more sessions and, therefore, can accommodate a greater number of participants than those of the single-day variety, which usually center on four or five prominent speakers. The biggest conferences tend to be sponsored by national or regional scholarly associations such as the MLA or the American Historical Association (AHA). Since they attract thousands of attendees and feature hundreds of speakers, they usually take place at large hotels and convention centers in major cities. Though such conferences can be overwhelming and are extremely impersonal, they have the advantage of being able to offer something of interest to just about everyone.

Except for the smallest gatherings, most conferences are built around multiple "sessions" focused on a unifying theme, each comprising a "panel" of speakers. While a plenary session may involve only one distinguished speaker, typical conference panels consist of an organizing chair or moderator and three speakers. In some fields, a "respondent" is permitted time to address the three papers and the connections between them. In the majority of cases, each session lasts about an hour and thirty minutes, which allows for a short introduction by the chair, three 20-minute presentations, and an approximately 30-minute long question-and-answer period. If panels run successively, they tend to attract larger audiences. If they run simultaneously, the size of the audience will vary drastically based on the prestige of the speakers and the attractiveness of the topic. Depending on the type of conference, the time of the session, and even factors as unpredictable as the weather, sessions will vary in shape and size.

As a speaker, you should think logically about how your session is likely to be run (it's okay to ask the chair ahead of time), but you also should be prepared for anything. In different situations, I have spoken to as few as five and as many as seventy-five audience members. I have been required to use a microphone and stand at a podium on a stage, and I have been asked to sit in a circle composed of armchair desks of the sort we all have in our classrooms. I have been asked to limit some presentations to ten minutes, and I have been permitted to speak at others for as many as forty-five minutes. I have given papers that attracted half a dozen questions and others that provoked only the most deafening silence. Your experiences will be no less diverse. As a rule, though, I would encourage you to plan your conference sessions according to what I call the 20/20 principle: that is, think of the average conference paper as a 20-minute presentation that you will deliver to about 20 audience members. You will find that this principle generally applies in the vast majority of conference scenarios.

Finally, most conferences offer a number of activities, cultural and otherwise, relevant to the particular meeting. Nearly all conferences include an overpriced luncheon or dinner, which, depending on the size and nature of the conference, may or may not be worth your time and money. Far more exciting and valuable are the cultural activities arranged by many conference organizers. A three-day drama conference might commence with an opening-night performance of *Pygmalion*. At a Jackson Pollock conference in New York City, you might be offered discount tickets to visit MOMA. Or as an attendee of a one-day conference on Spanish colonialism, you might be permitted free of charge to visit a local museum of Native-American history. Regarding such activities, pursue whatever you think might be useful or enjoyable.

WHY, WHEN, WHERE, AND HOW MANY?

When considering how many conferences to attend prior to graduation, you should remember that $1.25 and ten conference credits on your CV will get you a steaming cup of coffee. Conference presentations are not substitutes for publications, and they are worth very little in and of themselves. In other words, no Ph.D. has ever been hired because he happened to present papers at ten conferences. As a demonstration of scholarly activity supplementing a solid publication record, however, a strong list of conference presentations will no doubt strengthen your CV. So where does this information leave us? Why should I attend conferences? Which ones? How many?

Why You Should Attend

I'll never forget my excitement, as a second-year MA student, upon receiving my first conference paper acceptance letter. I sat down at my desk and enthusiastically whipped up a ten-page paper to end all conference papers. For weeks I revised and recited. Several months later I hopped a flight to Edmonton where I presented the paper to a small group of audience members. Now, 10 years later, the conference paper is still in my possession. Unfortunately, it's still ten pages long, and its content has never been published in any form. The idea underlying the paper is not half bad, actually. In fact, if I had the time or the inclination, I could probably turn it into a publication, but this would require tearing up the ten pages and starting from scratch.

At the time I wrote the paper, I had absolutely no idea what I was doing. I knew that I wanted to present a paper at a conference because this was what I saw my professors and advanced graduate-student colleagues doing. I had no understanding, though, why they were doing it. If I had asked, someone surely would have told me that I should think of conference presentations as drafts of papers with far more glorious futures ahead of them. Such a response might have prevented me from writing my paper as if it were the thing itself.

There are five good reasons to attend conferences, all of which have to do with the *future fate* of what you will present there:

1. First and foremost, you should attend because doing so may lead to the publication of ongoing research. A presentation to your colleagues will give you a sense of how well your ideas are likely to be received by a journal's readers. Their feedback should help you to identify the strengths and weaknesses of your paper, determine the adequacy of your bibliography (auditors love to suggest other works "you *really* should read"), and gain a sense of what needs to be elaborated, added, or cut from the paper. Think of the best conference experiences as opportunities for vetting your ideas.
2. Next, you should attend conferences because they are excellent venues for *advertising* your ongoing research. Especially at the dissertation, job hunting, and book marketing stages, it will be quite important for you to associate your name with a specific project or topic. Ideally you will create a situation wherein people thinking of a particular research topic—for example, eroticism in witchcraft rituals—will think specifically of your name. Conference activity should be stepped up, therefore, at periods when longer research projects are close to completion.
3. Attend conferences because of the networking opportunities they present. Scenarios in which networking is appropriate are far too numerous to record here, but consider the following example.

Let's say you've chosen to write your dissertation on the erotics of sixteenth-century English witchcraft rituals. While no one has studied this particular aspect of the rituals (they have, of course), Sir Genius Johnson is widely known as *the* expert on English witchcraft rituals. Having him on your dissertation committee would be quite a boon—a real endorsement of your work. You decide to submit an abstract for a panel he happens to be chairing. Or, you simply decide to attend a talk he will be giving at a local conference. When he sees your abstract or hears about your idea, he seems genuinely interested and asks you to keep in touch as the project continues to develop. Now you have established an acquaintanceship that might lead to bigger things.

4. When appropriate, attend conferences in order to pitch your research to editors. Most large conferences attract editors of major university and trade presses, who recognize conferences as ideal places to advertise their books and journals, to assess what's hot in the scholarly world, and to meet prospective authors. Especially when you are ready to begin revising your dissertation for publication as a book, conversations with editors may prove vital to your success.
5. Finally, attend conferences in order to immerse yourself in the professional culture of academe. While this advice may seem overly vague in comparison with points 1–4, it should be taken no less seriously. Conferences will teach you a good deal about the inner workings of your discipline, especially those pertaining to the publishing world. Any experiences you accumulate are likely to be useful down the road, which takes us back to the beginning: if nothing else, my first conference experience was useful insofar as it taught me *what not to do* the next time around.

When You Should Attend

Think about attending your first conference near the end of your second year or during any point in the third year. By this time, you will have enough experience writing seminar papers to understand the basics of academic research and writing, and you also will have a number of longer research projects from which to draw ideas and subject matter. There is no point in conferencing too early, since underdeveloped confidence and experience may lead to problems that should be avoided; conversely, there isn't much point in waiting too long to attend a conference since your confidence and understanding of the discipline will grow with each conference experience.

Which Conferences You Should Attend

Unless you are doing it solely to build confidence and have no plan whatsoever to record it on your CV, do not waste your time attending a graduate student conference. Know that you are more than capable of succeeding at a "real" conference and seek out the superior benefits of attending one. Know also that a job application boasting of a presentation at a graduate student conference will strike many search committee members as rather pathetic. Before you submit a paper, you might benefit from attending a conference either at your own or a nearby university. Seeing what goes on there will make you more confident about your ability to participate. Once you decide you are ready, submit an abstract or a paper for a regional (or a relatively small) conference. Once you have a presentation or two under your belt, be more ambitious and shoot for acceptances at national or international conferences. Talk to your advisors about those five conferences that are most important in your field, and make it a point to give a paper at one or more of them prior to graduation.

How Many Papers You Should Give

As in publishing, quality is always more important than quantity. Simply remember that two conference presentations that develop eventually into publications are worth far more than five or more presentations that don't. Stay focused on what you wish to gain from conference papers over the long haul. Avoid getting bogged down in numbers.

How to Apply for a Conference

The Selection Process

Sometimes a conference announcement will inspire you to submit—or generate from scratch—a paper or an abstract. At other times you will shop for a conference in order to submit a preexisting paper. In either case, the application process will be roughly the same. The first step you must take is to become aware of your options. Scholars now benefit from many useful websites that announce submission deadlines for upcoming conferences according to subject, topic, and discipline. More tried and true methods for discovering upcoming conferences and submission deadlines include checking the "Calls for Papers" regularly printed in field-specific journals, becoming a member in professional associations that sponsor annual conferences, and discussing what's available with advisors who are active conference-goers.

CFP: Milton in Modern Popular Culture (MLA '03; 3/15/03; 12/27–30/03)

Milton and Modern Popular Culture: An MLA panel sponsored by the Milton Society of America exploring modern pop-cultural appropriations and engagements with Milton's poetry and prose. Co-chairs Laura Lunger Knoppers and Gregory Colón Semenza are interested in essays that consider the presence of Milton in film, television, advertising, rock music, popular technologies, and other popular media. Essays on the uses of such media in the Milton classroom are welcome. Please send abstracts of 500 words or complete, 20-minute papers to either of the following two addresses by March 15, 2003.

Figure 9.1 Sample call for papers

Let's consider a hypothetical scenario by which you might select an appropriate conference. In one of your first seminars, your professor mentioned a University of Pennsylvania website (http://www.english.upenn.edu/CFP/), which posts Calls for Papers related to all subjects of English studies. You immediately bookmarked the site and have made it a point to check about once a month the "recent messages" for British studies. One day a particular announcement catches your eye. Figure 9.1 reproduces a call for papers as it would appear on the website.

Notice that the CFP announces both the dates by which submissions must be received and the dates during which the actual conference will be held. The announcement clearly indicates the type of paper the panel chair will be happy to receive. Since you wrote a very solid paper on *Blade Runner* just last semester, this conference offers you a good opportunity to begin reworking it for publication. Plus, it's nearby. You decide to work up the requested one-page abstract.

Generating an Abstract

An "abstract" is simply a summary of a text. For conference applications, you will write two kinds of abstract. The first—usually for a paper that you have not yet written—takes the form of a more overt proposal: "In this paper, I will explore X". The second—usually for a paper you've already written in one form or another—tends to be more argumentative and may even present conclusions. As the two abstracts would suggest, conference proposals share several common rhetorical features, though they may order those parts differently. First, as both examples in the appendix would suggest, an abstract usually offers some sort of historical or scholarly context out of which the eventual argument will grow. If the essay is nonhistorical, you may begin by covering a scholarly debate. If it is historical, you'll want to put the historical facts on the table fairly early on. Second, give a sense of how you read the text, other cultural artifact, or the historical context differently from previous investigators—what we called in chapter 5 the "situating move" (see pp. 105–07). Tell us what you are contributing to existing scholarly discourses. Finally, deliver your argument or, if you haven't yet written the paper, your hypothesis, along with a sense of its implications for current and future scholarship.

Always be careful to follow a chair's directions to a tee. If he requests a one-page single-spaced abstract, do not send a longer document. Like typos, ugly print-outs, and other signs of unprofessionalism, not following directions gives conference chairs convenient reasons to select other people's papers.

Writing the Cover Letter

As always, cover letters should be simple. Explain that you are responding to the Call for Papers. Give the title of your paper with no more than a one-sentence description of its content. Make sure you provide contact information so that the chair can easily get back to you, and thank the addressee for his time. Pages 303–04 offer an example of a typical cover letter for submitting a conference abstract. Once you put the abstract in the mail, be patient. It may be months before you hear anything. In most cases, panel chairs alone decide which papers to accept. A CFP for a local conference may attract only five or six proposals. A CFP for an international or national conference might attract 50 or more. The duration of your wait, therefore, will depend largely on the size of the conference at which you hope to present your paper.

PREPARING THE PRESENTATION

Once your paper has been accepted, it's time to begin shaping the actual presentation. The amount of work you'll need to do will depend largely on whether you're condensing and reworking a seminar paper (highly recommended) or creating a presentation from scratch. In either case, while writing the paper try to keep your mind focused on the odd and contradictory form of a typical humanities presentation: on the one hand, we usually read our presentations, as opposed to memorizing them or guiding them with software such as PowerPoint; on the other hand, our audience has none of the benefits of typical readers, such as the ability to alter pace, skip tedious material, or review difficult passages. Good presenters think carefully about the difficult position of their audience members, and they adjust their presentations accordingly.

None of this will be shockingly new to you. Effective conference presentations draw on three skills developed early on in graduate school: oral reporting, teaching, and seminar paper writing. In what follows, I offer advice about how to prepare conference presentations by drawing specifically on these familiar skills. If your experience in one or more of these areas has been limited, you'll simply need to spend a bit more time thinking about how to address certain issues covered below.

The Oral Report as Practice

Presumably, if you're planning to attend a conference, you've already read several papers in front of a crowd. Oral reports are standard assignments in humanities seminars, especially the ubiquitous "Review of Literature/Criticism" or "Annotated Bibliography" reports that we all come to know and hate. But even such seemingly tedious assignments as these pay infinite dividends later on when we arrive at our first conferences. Both the oral reports that we deliver in seminars and those we hear others deliver teach us a great deal about the attributes of a successful presentation. In relation to conference papers specifically, we can break down these attributes into three equally important categories:

Presence: Memorable speakers are never sheepish. Nor are they difficult to hear or understand. Not every person can simply light up a room by entering it, but most of us can practice presentational behavior that keeps attention adequately focused on us. Your eyes are the chief magnet, of course, and so it's extremely important to look up from your paper now and then to meet the various sections of the room. Consider your own experience and take note of your behavior the next time you attend a talk of any sort: if the speaker refuses at any point to look up from his paper or if he looks only at one side of the room, you will be infinitely more likely to allow your eyes to wander; you may even roll your eyes, yawn, or sigh loudly. If the speaker looks at you occasionally, though, you will be very unlikely to do any of these things. The point is not merely that such behavior is rude and, therefore, to be avoided. The point, rather, is that occasional eye contact will keep auditors focused on you and what you're saying. If you have a weak voice or tend to mumble, practice delivering your presentations aloud to partners and friends. Train yourself to hear the new, loud voice you will bring to the conference. Nothing puts an audience to sleep more quickly than a soft voice.

Pacing and Time Management: An important fact worth stressing is that, so far as I know, no one has ever complained about a scholarly presentation being too short. One major advantage of a shorter paper is that it allows you to establish a perfect speaking pace. Ask experienced conference-goers about the average length of their typical 20-minute presentations, and they will give you answers ranging from 8 to 12 pages. The range is accounted for by the fact that whereas the 12-pagers fly through their presentations, the 8-pagers may actually read too slowly. As a rule, the more nervous you are, the more quickly your presentation will go. As a graduate instructor, I've noticed that first-year MA students tend to speak at a much faster pace than more experienced students. Make it your goal to establish a pace that allows you to speak clearly, to pause after important points, to intersperse impromptu comments where appropriate, and to emphasize specific words or passages. Practice reading your presentation aloud with the clock running. Whatever you do, make sure that you do not exceed the time limit mandated by your panel chair or conference organizer. At best, an excessively long paper will succeed in annoying everyone in the room. At worst, the chair will cut off your presentation before you are able to conclude, a humiliating situation that I have witnessed on two occasions. A good rule to follow: in practicing for a 20-minute presentation, try to time out consistently at 19 minutes.

Diplomacy and Collegiality: If "Literature Review"-type reports teach us anything, it's how to discuss other scholars' work in more or less appropriate ways. Since as always, you will need to situate your work in relation to previous scholarship (see pp. 105–07), the skills you've already learned in your seminars will prove invaluable at conferences. Just avoid falling into the trap of assuming that a harsh critique of an author demonstrates intellectual rigor. Focus less on what you perceive to be flaws and more on what you perceive to be strengths, and highlight what your audience needs to know to be able to measure your contribution. An overly belligerent or negative tone will send a clear message to your audience about how you wish to define the terms of the conversation; if you suggest that scholarly conversations are to be combative rather than collegial and respectful, you should expect to be attacked in turn during the question and answer period. Work instead to establish the sort of constructive tone and terminology that you would like others to employ when confronting your research.

Teaching as Practice

One *could* think of a conference presentation solely from the perspective of the presenter. I would caution you to avoid doing so. Understanding how teaching is analogous to the presentation of a conference paper nicely emphasizes the fact that such presentations need to be considered from the perspective of the audience. The point of delivering a presentation is not merely to display one's learning, in other words, but to alter or enhance the ways in which the subject matter is perceived and understood by an audience. With this point in mind, you have much to gain from drawing on skills you've likely been practicing in the classroom.

Enthusiasm

As you know from your role as both a student and teacher, there is no adequate substitute for passion in learning situations. Because so many academic talks are dry and, well, *academic*, lively and enthusiastic speakers tend to separate themselves from the crowd. Find in your paper what you're most excited about and work to convey this excitement to your audience. Create a sense of exigency about what you're doing in order to keep your audience alert and focused on what you have to say.

Organization

Though we all can probably cite exceptions to the rule, good teachers tend to be organized, primarily in the sense that they help us to know where we've been, where we are, and where we happen to be going. One major difference between a conference paper and an article is that the former should be much more explicitly architectural and aware of its structure. Since your audience will miss out on the benefits of being able to control their "reading" of your paper, you should go out of your way to help them to follow your argument and your methodology. Whereas metawriting ("In this paper, I will do X") *can be* awkward or even annoying in a seminar paper or article (usually, it's very effective), it is almost always appreciated in a conference setting. Especially in your introduction, make it a point to announce your objectives, as shown in figure 9.2.

What I will focus on in this talk is the complex manner in which the category, youth, is constructed by *The Animated Tales* and, more specifically, the implications of this construction for the target audience of 10–15 year olds. Looking closely at the manner in which the *Tales* translate Shakespeare's plays into short children's films yields interesting insights into the ways that kids are understood and reproduced by a Shakespeare industry increasingly influenced by corporate ideals and objectives. More importantly, to the degree that such corporate ideals run contrary to the ideals of a critical and self-conscious democratic society, such an analysis will hopefully suggest the need to critique and actively challenge overly simple, corporate-based notions of Shakespeare as a practical tool for the socialization of children.

Figure 9.2 Sample conference paper introduction

In addition to announcing your goals, stop now and then to reiterate important points and to highlight rhetorical transitions on which your arguments hinges. In concluding your presentation, provide a brief summary of the paper, hitting on the major points and the relevant terms so that they are fresh in your audience's mind. You'll find that questions will be more specific and helpful as a result.

The Seminar Paper as Practice

The most important preparation for conference presentations will come from seminar paper writing. Though the tone *may be* more casual and the style more rhetorically affected, a typical conference presentation looks a lot like a short scholarly article without footnotes. Since we've already discussed the form of a seminar paper in a previous chapter, briefly consider here how the various parts might differ in a conference paper format.

The Argument

As in a seminar paper, you'll want to offer your audience a clear, provocative claim. Unlike claims that appear in seminar papers, though, those in conference presentations should be introduced and reiterated throughout the paper in the form of direct, even obvious formulations. Consider the same argument as it might appear first in a seminar paper and then in a conference paper, as shown in figure 9.3.

Nothing would necessarily prevent an author from using the first introduction in a conference presentation, but the audience would feel less of a connection to a speaker who chose to do so. Obviously the second introduction—a useful example of how much more liberally metawriting can be employed in conference settings—would be wholly inappropriate in an article. Humor, however weak it may be, is also more appropriate for a conference setting, as is the slightly more colloquial voice of the second introduction. Finally, the second example demonstrates how presenters often work harder to contextualize their local arguments in relation to their larger projects—a move that helps to clarify for an audience the significance of that local argument and also allows the author to advertise his research more effectively.

Because a conference paper can be understood as a work in progress, you might also choose to go further out on a limb, when constructing your argument, than you might ordinarily do. That is, you might wish to state your claim in a slightly more radical way in order to draw in your audience. A presenter beginning his paper with the bold claim that "God has been dead since at least the eighteenth

Example 1: Seminar Paper or Article

Izaak Walton's *Compleat Angler* is certainly the most successful sporting treatise ever written. Never out of print since the first edition of 1653, the *Angler* ranks only behind the Bible and the Book of Common Prayer as the most frequently published work in the English language. Traditionally characterized as a simple pastoral dialogue by an equally simple, even accidental, author, the *Angler* has more recently been viewed as an allegorical protest against the precision of the Interregnum. While historians and literary critics have helped to reveal the *Angler*'s general political context, however, no scholar has done justice to Walton's complex and highly specific engagement of official Interregnum policies regarding sports and pastimes. Most recent work has attempted to reconcile a traditional portrait of Walton as an innocuous, simple-minded countryman with a growing awareness of the political suggestiveness of his literary masterpiece. But Walton does more than passively evoke the mythological image of a pre-Interregnum golden age; in fact, he uses sport quite deliberately and systematically to critique contemporary laws proscribing communal recreations.

Example 2: Conference Presentation

When I was invited to talk today on any aspect of my current research, it occurred to me that my choice of a subject *should*, in fact, be rather easy. My research has involved examining sports and sporting events in Early Modern England—including the decidedly ungraceful sports of football, shin-kicking, wrestling, and bull baiting—sports that in their raw violence and capacity for excess have the power to shock the modern imagination. What I *should* do, it seemed obvious, was choose a sport with the power to shock, one that my audience simply *could not* resist. So after long and rather meticulous consideration, the right sport to discuss became obvious to me: fishing.

Seriously speaking, though, there are *three* things in particular about the subject of Early Modern sport that I'd like you to take away from this room. First of all, I want to show you how the inherently contradictory nature of sport—its ability to emblematize extreme states of order and disorder—helps to explain its political power and complexity as a metaphorical tool. I provide a detailed *example* of sport as such by demonstrating how Izaak Walton used angling to critique the legislative policies of the Interregnum . . .

Figure 9.3 Comparison of conference and seminar paper arguments

century" would likely perk up more heads than one beginning, "One can argue that the major Enlightenment thinkers succeeded in challenging many Europeans' absolute faith in the idea of an omnipotent and omniscient deity." Offsetting the radical formulation with a disclaimer reminding audience members that your paper is part of a "work in progress" helps to explain, even justify, your willingness to push things further than usual. It goes without saying that the more provocative your paper, the more likely is an audience to remember it and offer useful feedback on it. You should never "invent" an argument merely for rhetorical effect and you should always be sure to back up your claims later in your paper, but you should also do what you can to make people notice the argument that you are presenting.

The Situating Move

Generally speaking, your review of literature should be much more concise in conference presentations than it is in seminar papers. To the degree that it's useful, try to summarize movements or trends in the scholarship rather than the arguments of individual researchers. Avoid dwelling on obscure scholarship; whether it's fair or not, you'll get more mileage out of a reference to one of Richard Rorty's obliquely relevant essays than you will to a more obscure philosopher's directly applicable one. Audiences know who Rorty is and they may know his work quite well, so your reference will serve to keep them tuned in to what you're saying. Most of all, be generous to previous scholars, as you will want future scholars to be generous unto you.

The Evidence

Most seminar papers build arguments cumulatively, on multiple forms of evidence and numerous examples. Since conference papers are less than half the size of a typical seminar paper, however, you will have a difficult decision to make: should I present all of the examples and offer less commentary on each, or should I present fewer but more detailed examples? While different situations demand different responses, the second answer is, nine times out of ten, the better one. First, a conference paper format allows you to explain to your audience what they already know: that time is limited, which also limits the range of issues you can discuss. If you happened to select for presentation only three of the fifteen examples featured in the seminar paper version, you can simply tell the audience that you've done so and offer to discuss other examples during the question and answer period. Second, the detail you lavish onto a few, well-chosen examples, will do far more to help you build a case than will multiple, shallow readings of evidence, which an audience will try to deconstruct before they even leave the room. Finally, your audience will appreciate the limited range of topics you're choosing to discuss; instead of trying to keep up with you as you move through 15 different forms of evidence, they will be more than capable of balancing in their minds three different ideas, which, together, demonstrate the persuasiveness of your argument.

In thinking about which examples or forms of evidence to present, consider first which ones are most persuasive; second, which ones are the most vivid and poignant; and finally, which ones are likely to be most familiar to your audience. Remember that whereas you can force readers of print materials to work in rather demanding ways, you'll need to make your conference presentations as easy to follow as you possibly can.

In a twenty-minute presentation, you might dedicate a page or two to introducing your argument, a page or two to situating it in relation to previous scholarship, and six to eight pages to analyzing a few well-chosen proofs of your argument's persuasiveness. While preparing all three parts, keep in mind the importance of clean and clear transitions, and remember that clarity is the basic characteristic of a solid presentation.

At the Conference

By finishing your paper ahead of time, you'll be able to participate in, and perhaps even enjoy, the actual conference. Here are some suggestions for making the most of your trip.

What to Bring

Clothing at conferences is no more formal than it is on campus. Dress as you would to teach. Bring one nice outfit for each day you plan to attend sessions, and take along a travel iron if the hotel doesn't provide one in your room. Casual clothes are a good idea too since you'll probably want to see the host site or even get in your exercise. My favorite thing about conferences is the opportunities they allow for travel: from conferencing alone, I have been able to visit most major American cities, a good chunk of Canada, and several places in Europe and the Caribbean. I always bring a backpack, where I can keep such items as my bottled water, sunglasses, and even a travel guide. My advice is to get out of the conference now and then.

Okay, back to work. Bring two hard copies of your paper. Keep them in separate places. If you check luggage, place one copy in the checked bag, and carry the other one on your person. I always bring with me an electronic copy on a disk or e-mail one to myself just in case I decide to make changes to the version I read on the plane or in the hotel room before my session.

Bring a folder with all relevant information about the conference, including your hotel information, the conference program, a notepad and pen, receipt of your registration payment, and so on. I also suggest that you type up a brief (three to four sentences) biographical statement. Chairs like to personalize their introductions, but they rarely have the wherewithal to research you ahead of time. Nine times out of ten they will ask you to write up something quickly when you enter the room for your session. Prepare the biography ahead of time so that you will be introduced in a manner that you feel comfortable with. If your presentation comes from a seminar paper or dissertation chapter, bring the longer version so you can read it the night before; you'll want additional information and the complete review of literature fresh in your mind for the question and answer period. If you plan to pass out handouts, make twice as many photocopies as you anticipate needing.

Finally, if you have the equivalent of business cards, bring a few along so that you can give them out when appropriate.

Checking In

As soon as you check in to the hotel, head down to the conference site to register. Most conference organizers prepare packets for registered participants that you'll want to check out as soon as possible. Sometimes session times are altered so be sure that you know exactly when you are expected to show up for work. Most registration packets include name-tags, conference programs, information about the host city, and other pertinent information. I strongly recommend that you pay the registration fee ahead of time since most conferences will cost you more money if you pay on-site.

Find out exactly where your presentation will take place and find the room ahead of time. Count the number of seats, note whether or not there is a microphone or a podium or an overhead projector, and check out more basic things like temperature and lighting. Visualize what you will face when the time for your presentation arrives, and read your paper aloud at least one more time with the image of the room in your mind.

The Conference Begins

At larger, more prestigious conferences, the first thing you will notice is that pretentiousness reigns. People will glance at your name-tag, see that you are nobody important, and shuffle along. Be prepared for such rude behavior and learn to laugh it off. On the other hand, you'll probably be surprised by the number of strangers who will suddenly begin chatting with you. This is no time to be a wall flower. Conjure up whatever enthusiasm and charm you can find inside, and socialize with your colleagues. Especially for graduate students, who very well might see some of these people in interview rooms later on, such networking may prove very beneficial. Remember also that rank and the usual academic hierarchies apply less in conference settings than they do on campuses; that is, when socializing, you should act like a colleague and avoid seeming overly deferential. And if you should meet that superstar who has influenced your work, avoid gushing or acting in a sycophantic way, which will turn off everyone around you, including the superstar. Networking is, of course, a crucial component of conference-going, and it is likely to cause you more stress than even the presentation itself. I hate clichés, but on this one, I agree you should simply be yourself. Acting like someone else—*acting*, in general—will only make you unlikable.

Should you have the opportunity to meet an editor—of a journal, a volume of essays, or an academic press—let them know what you're working on. Again, act like a colleague and avoid offering voluntarily the irrelevant information that you are only a graduate student. Editors are more likely to care about your work than your status. Often they will offer you useful information about how to make a particular project more marketable, especially at the dissertation stage. Occasionally, your conversations with an editor will lead to bigger and better things such as publication opportunities. At least three of my own publications have grown out of conversations with editors at conferences. Networking is sometimes tedious business, and some people dread this aspect of conferences for perfectly good reasons. But it is worth your time and energy to pursue relationships that are likely to serve you in various ways throughout your career.

As an audience member in attendance of conference sessions, you will learn a great deal about the likely dynamic of your own upcoming session and about conference presentations in general. Focus carefully on what works for you in the presentation and what doesn't, and tune in to audience behavior and body language as well. Take additional notes on the conduct of the chair, since you may wish to organize a panel sometime prior to graduation. All of this information will undoubtedly serve you at various times both during and after the conference is over. So unless you happen to be presenting in the very first session of a conference, make it a point to attend at least one session prior to your own presentation.

The Presentation and Q&A

Show up about ten to fifteen minutes prior to the start of your session. You won't necessarily have an opportunity to meet the conference chair (and other panel members) before the actual session so allow yourself some extra time for introductions and chitchat. If you've brought a biographical statement and the situation seems appropriate, give it to the chair now. If you need to set up an overhead, cue a film clip, or pass out handouts, do so now. As the session time approaches, take a deep breath, relax, and realize with confidence that you are more than ready to go.

Sometimes a chair will allow questions immediately after individual presentations, but usually questions will be saved for a period of time after the entire panel has presented. Your experience delivering oral reports has undoubtedly made you adept at answering hard questions, and you should try, simply as a matter of strategy, to anticipate the worst before facing a conference audience. You will almost certainly be surprised by the collegiality of most of your audience members. If there happened to be a "right" way to answer questions, I would happily tell you what it is. But since different speakers handle questions in so many different, wonderful ways, I'll focus instead on a much easier subject: the *wrong* way to answer questions.

Never Pretend to Know The Answers When You Don't

Not only will your audience recognize when you're fudging, but you will also undermine the persuasiveness of your entire presentation by seeming dishonest. If you don't know the answer to a question, simply say so: "You know, your question reminds me of how much I still have to research in the area of cognitive theory. Since this is a work in progress, I just don't know yet how exactly to answer the question. I'd certainly be interested in hearing your sense of things." Notice how such an answer not only calls attention to the idea of a conference presentation as an incomplete and evolving work, but it also pacifies the questioner by empowering him to reverse roles and become the teacher for a minute.

Next, Never Pretend to Understand a Poorly Articulated Question

If you are completely lost, ask that the audience member repeat the question. If the problem is related to confusing or complex terminology, seek to clarify the matter before answering. Let's say that your audience member asks you about the influence of French theory on your understanding of a particular text. Since "French theory" is a fairly vague term, you might begin by working to establish that you are on the same page with your examiner. If you simply don't understand what the person means by "French theory," ask him, "would you mind clarifying what you mean by 'French theory'?" Or if you have a pretty good sense of the matter, you might approach the problem in the following way: "By 'French theory,' I take you to mean X, Y, and Z. Is this correct?" Once the term has been defined, you can move onto your response.

Never Blow Off a Question or a Comment

You will probably be surprised by how many audience members raise their hands simply to declare their opinion on a subject your presentation happened only to touch on. The individual has no intention of asking you a question. Probably, he just wants to hear himself speak. Assume, first of all, that such behavior annoys others in the room as much as it annoys you. But show yourself to be diplomatic by responding in some way or other, even if only to acknowledge the validity of the audience member's opinion. By doing so, you will soothe the individual's ego, and you will impress the other audience members with your ability to remain professional and collegial under strained circumstances.

Most of All, Avoid Seeming Condescending

Treat every question as though it's the best idea you've ever encountered, and you will earn the respect and good will of the audience members, who will continue to think well of you long after the conference. If possible tie your responses into ideas covered by the other presenters on your panel, which will show that you paid attention to their work as well. The question and answer period, when conducted effectively, should feel like a conversation between colleagues, not an interrogation or a lecture. If you are like most people, you will enjoy witnessing and participating in such conversations as much as any other part of the experience.

CONCLUSION

I won't deny it: I've attended conferences that were so boring, the paper topics so cliché, and the people so pompous and competitive that I happily spent more time in my hotel room watching ESPN than I did attending sessions. At times, the audiences in attendance of my talks have been so small that I have felt lucky just to answer one question about my paper. At other times, I have endured listening to multiple audience members going on and on about their "sense of the matter," completely uninterested in engaging the panelists' papers. There is no question: at their worst, conferences are an utter waste of time (though tourism can make the worst of them tolerable).

At their best, though, conferences can benefit you and your university in various ways. On a personal level, you will build important professional connections, learn a good deal about publishing practices, discover hot topics in your field, and, hopefully, receive valuable feedback on your own research that may lead to publication. The oral presentation and the Q&A component constitute excellent preparations for job market scenarios, as does the experience of socializing with your peers at other universities. Your university will benefit from supporting your participation at national conferences since it looks good when you look good. In short, by seizing the opportunity to attend an appropriate number of sessions, present your research, and enjoy a few well-chosen local activities, you will find many conferences to be illuminating and enriching experiences.

CHAPTER 10

PUBLISHING

Before mailing out your first application for a tenure-track job, set the goal of publishing (i.e., having *accepted* for publication) at least two article-length pieces; make sure that one comes from your dissertation. While such advice may seem unrealistic, even terrifying, to you at this stage of your career, careful long-term planning and an informed approach to seminar-paper writing can lead quite naturally to the generation of publishable material. The key is to avoid getting too far ahead of yourself since the pressure to publish can sometimes be paralyzing for inexperienced students and professors. Publishing success, on the other hand, can lead not only to exciting job prospects but also the personal satisfaction of knowing you have reached the pinnacle of accomplishment in your field. No one ever forgets the first time.

Generally speaking, the publishing process consists of three distinct phases: the first phase, of course, involves the actual creation of publishable material. If you have not yet read chapter 5, stop and do so now, since it deals extensively with *how to write* a publishable piece, and then return to this chapter. The second phase has to do with the actual mechanics of publishing, the complex procedure of selecting an appropriate outlet and convincing its board to publish your work. The final phase is dedicated to preparing an accepted piece for its permanent appearance in print (or, increasingly, online). Focusing mainly on the second and third phases of the publishing process, this chapter addresses the following subjects:

- The forms of publication in the humanities
- The selection of an appropriate publisher
- The communication and correspondence processes
- The long wait for a response
- The different types of editorial response
- Final revisions and proofing

THE TWO-ARTICLE GOAL

There is no reason any more to hire a Ph.D. who has yet to prove she can publish. Because most moderately desirable jobs in the humanities will attract 100 or more applicants, the chair of any search committee can rest assured that many of them will have published one or more articles. Since many, if not most, of these hiring schools will require from the applicant a published book (or the equivalent in articles) for tenure, the chair of the search committee would also be wise to demand an excellent and highly marketable dissertation. Since the appearance of a dissertation chapter in a peer-reviewed journal is something like proof of your dissertation's legitimacy and marketability, your number one priority should be to publish a central chapter from that project. In order to preempt the possibility of a search committee regarding your single publication as a fluke, shoot for two publications prior to sending off those job applications. Your second article can also be from the dissertation, but you will probably benefit more by publishing on a different topic in your field, which demonstrates your range. Again, avoid publishing more than two chapters or about 25 percent of your dissertation material.

Now before you panic, be careful to keep the implications of this two-publication goal in proper perspective. Even if you finish your Ph.D. on time, you will still have six years to get two seminar-paper-length pieces accepted for publication. In your course-work alone, you will write approximately twenty such papers, and you will add another five or six at the dissertation stage. Out of roughly twenty-five papers, then, two will need to be polished enough for approval by a respectable journal's readers. Remember that most graduate students earn their first publication credit while an ABD, which means there is plenty of time to "master" academic writing before sending anything off. While MA students will benefit immensely by establishing professional goals early in their graduate careers, it might be detrimental for some of them to rush what will very likely take many years to accomplish.

As my wording has suggested, the "acceptance" of your work for publication is more than adequate for a competitive run on the job market and may even be preferable to having one or more pieces already in print. Tenure committees are more likely to count articles toward tenure if they bear the name of the tenure-granting institution; since an accepted piece is likely to be revised, expanded, and proofed after a contract has been signed, an assistant professor can note in an article the new affiliation with the hiring university. "Old" articles, on the other hand, will help you to get hired but probably will not count toward tenure. In an ideal situation, therefore, a job candidate will enter the market with two recently accepted, peer-reviewed articles. You might even establish as part of your contract negotiations that such articles will count for tenure and/or merit raises (though you should recognize that few universities will want to count them).

The Forms of Publication

While we have been stressing the importance of the peer-reviewed article as an indicator of one's scholarly credentials, several other forms of publication are expected of humanities scholars. A book review or a chapter in an edited collection will never be regarded as the equivalent of a peer-reviewed article, but such publications will certainly strengthen any CV. Several common forms of publication in the humanities are listed and described below in descending order from the most to the least important.

The Peer-Reviewed Monograph

In 2002, then president of the MLA, Stephen Greenblatt addressed a personal letter to all members about a "serious problem in the publishing of scholarly books." Because of considerable budget cuts to university and academic presses, which would prevent many younger scholars from publishing their first book prior to tenure review, Greenblatt implores departments to discover alternative means of evaluating scholarly productivity:

> We could try to persuade departments and universities to change their expectations for tenure reviews: after all, these expectations are, for the most part, set by us and not by administrators. The book has only fairly recently emerged as the sine qua non and even now is not uniformly the requirement in all academic fields.[1]

Greenblatt is referring, of course, to the serious damage to academic presses caused by regular slashes to university library budgets over the past three decades. Kathryn Hume explains why such cuts have so impacted the academic publishing industy:

> In 1970, the standing library order for books from prestigious American academic presses was over eight hundred copies. . . . Around the year 2000, the standing order is about one hundred and seventy copies. Producing an ordinary monograph now costs $8–$10,000 in

> direct costs, and $16–$20,000 in indirect costs. . . . To get the $30,000 back, a press would have to charge over $170 per book.[2]

According to Hume, most presses lose money even when they sell books for $50 or more per copy. So how should a current graduate student and future professor respond to these book publishing problems, which she may very well face in the next few years?

Despite the current crisis in the academic publishing industry, the scholarly "book" remains the most prestigious form of publication in the humanities. Unfortunately, there are few indications that competitive research universities will stop demanding it as the chief criterion for tenure and promotion decisions. My advice, therefore, as chapter 8 suggests (see pp. 170–71), is that you write a dissertation that seems likely to become a book, and that you assume you will need to publish that book for tenure, which may or may not prove to be the case down the road. Though it is important for you to recognize now the centrality of the peer-reviewed monograph in humanities disciplines—as well as the economic crisis that threatens to make it extinct—you need not worry about trying to publish a book prior to graduation. Book publication will become one of your primary goals when you are hired on the tenure track.

The Peer-Reviewed Edited (Book) Collection of Essays

Surely some academics would argue that the edited collection of essays belongs in a lower spot on this list. A strong collection, however, is capable of transforming scholarship in a particular field, and it can also enhance considerably the reputations of both its editor and its editor's department. Solid collections of essays require a tremendous amount of work from editors, who must define the book's topic, coordinate an appropriate list of contributors, edit and copyedit the contributors' essays, write an introduction, and, often, contribute an original essay to the volume. While individual departments are responsible for judging the relative value of a particular collection of essays, it goes without saying that a cutting-edge collection has greater potential for impacting a scholarly field than a journal article. You should not think about editing a collection, however, until after you have completed your dissertation and turned it into a book.

The Peer-Reviewed Journal Article

Still the most basic building block of any academic career, an article placed in a respected journal may be read by hundreds of scholars and has the ability to influence the author's reputation and the wider understanding of the subject she discusses. Later sections of this chapter will delve rather deeply into the process of publishing an article in a peer-reviewed journal.

Chapters in Collections of Essays

Some such essays are peer-reviewed by the editors themselves or by readers from a press still deciding whether or not to publish the collection. Since the assumption of some academics tends to be that chapters in collections are *not* peer-reviewed even though nearly all of those collections published by university presses *are* peer-reviewed, you should make absolutely clear (and be willing to show) that a particular essay has in fact been peer-reviewed if that is the case. There are several scenarios by which one of your essays might wind up in a collection of essays, and together they reveal the problems in trying to define what constitutes a peer-reviewed article: (1) You come across a Call for Papers relevant to a particular topic on which an editor is attempting to publish a collection. You send the editor (or editors) the piece, which is read and accepted. Is this a peer-reviewed article? One can certainly make the case. (2) Responding to a Call for Papers, you send an abstract or proposal for an essay to an editor or editors. They agree to include your essay in the collection on the basis of this abstract. In this case, you would have to stretch the facts to consider the piece a peer-reviewed article. (3) After delivering a conference paper, you are approached by an editor who asks whether you might be willing to include the piece in her forthcoming collection of essays. This should not be regarded as peer-reviewed. (4) Finally, you are commissioned (by phone, e-mail, letter, etc.) to write an essay for a particular collection of essays. In some cases, commissioned pieces should not be regarded as peer-reviewed. In many cases, though, essays commissioned by editors will have to be accepted by readers at the publishing press. For example, though I once was commissioned to write an essay by editors of an Oxford University Press volume on Renaissance drama, all of the essays were sent out to Oxford readers who judged their worthiness for inclusion in the volume. Because of the difficulties of defining peer-review in relation to collection contributions, you must assess rather carefully the process by which you have published any essay that appears in a collection. Talk to your department head about how you should refer to the piece on your CV. If you determine that you cannot describe the essay as having been peer-reviewed, be sure at least to make clear on your CV that the essay was commissioned if this was the case. Commissions suggest that you have already established a reputation in a particular field, and they are the next best thing to a peer-reviewed acceptance.

For the same reasons that a solid collection of essays is important, the inclusion of your essay in such a volume looks impressive and should be regarded as a sign of scholarly potential and excellence.

The Peer-Reviewed Note or Query

A former professor once explained to me that any idea worth publishing deserves more space than that afforded by a note. I disagree. While it's true that you should never settle for a note where an article is an option, one can easily imagine several situations in which the publication of a note is appropriate. In order to ensure that you are making the right decision to publish a note, seek advice from your advisors and experts in the field, since they may have ideas about how to expand a short piece into a more substantial one.

Book Reviews

The vast majority of book reviews are commissioned by journals so it is unlikely you will land many prior to the publication of your own first monograph. A very few respectable journals, however, read and sometimes publish un-commissioned reviews. While I would not recommend spending any significant amount of time trying to publish a book review, situations do arise in graduate school that lead students to publish reviews. Since I require that my graduate students review at least one recently published book related to the seminar I am then teaching, it makes sense for them to seek a home for the reviews upon which they have so diligently toiled. Usually at least one student each semester succeeds in placing a review in a solid journal.

A well-written book review displays a scholar's critical acumen and authoritative voice. You will learn a good deal about the art of book reviewing simply by reading published reviews in top journals. Like most articles, book reviews tend to be quite formulaic, moving from a description of the book's central argument to a discussion of individual chapters, to a discussion of the book's weaknesses, to a final paragraph that declares the book "a welcome addition" to the field. In writing reviews—whether for class or for publication—try to keep the following ideas in mind:

1. The first question readers will always ask after reading a review is "Do I now know what this book is about?" This sounds simple,

but you may be surprised by how many reviewers prattle on about minute details without ever conveying a sense of the book's major claims or how it is to be situated within previous scholarly discourses. Be sure to recognize not only what an author is arguing, but also the sorts of materials, evidence, and methodology she is using to make the argument.

2. Provide enough samples of the author's own language to do her justice. Be extremely careful when recasting an author's major points in your own language. There's a heavy burden of responsibility on every reviewer's shoulders.
3. Feel free to comment on matters such as style and issues pertaining to the book's apparatus (index, bibliography, etc.).
4. Attempt to balance the need to be fair and moderate with the expectation that you will point out the book's weaknesses. No book is perfect, and all of your readers know it, so unequivocal praise may actually irritate readers. On the other hand, whatever you do, don't be petty. Academe is a small world, and you do not want to offend colleagues who are likely to judge your work over the coming years.
5. Deal with the author's argument on its own terms. The fact that you don't like Marxism is not a good reason in and of itself to slam a Marxist approach. Seek out and try to understand the internal logic of every argument you encounter.
6. Regarding the cliché, final sentence about the book being a welcome addition to the field, think about how you might vary the idea. Think about what the cliché says implicitly and consider recasting the point in more local terms. For example, you may wish to emphasize in closing what happens to be the specific contribution of the book to the scholarly understanding of a focused research topic.

Other Forms of Publication

Humanities scholars also publish encyclopedia articles, op-ed pieces, conference proceedings, and any variety of non-peer-reviewed writings. Always remember that publications such as book reviews and encyclopedia entries can look good on a CV as indications of *additional* scholarly activity, but they should never be regarded as substitutes for or equals of peer-reviewed publications.

Publishing a Peer-Reviewed Article

From conception to actual printing, publishing an article is a several-year-long process. As with most things academic, the various steps involved in publishing an essay need to be learned; there is nothing obvious about any of them. Since chapter 5 outlines the initial phase of the article-writing process, the following material is focused on how you should proceed once you and your advisors have determined that an essay is ready to be submitted for publication.

Selecting a Journal

The first rule of publishing is that quality is far more important than quantity. Your job prospects and, indeed, your reputation will benefit more from one well-placed, excellently written article than three or four insignificant ones. Further, there is no point seeking publication in a mediocre journal unless you have first been rejected by all of the superior ones. The key is figuring out those journals that are superior and most appropriate for your piece.

Of course, by the time you are ready to submit something for publication, you will likely have a good sense of the top journals in your field. Course-work, exams, dissertation research, and most of all, your focused research on the piece, should all combine to suggest highly useful patterns in your mind: the most cutting-edge articles on Shakespeare tend to appear in *Shakespeare Quarterly*. More old-fashioned historical ones tend to appear in *Renaissance Quarterly*. New Historicist ones appear in *English Literary Renaissance*. Heavily theoretical ones in *Representations*. In some cases, though, a majority of Shakespeare articles on the particular topic being researched will happen to have appeared in *SEL*, which would make it a logical choice for an initial submission. You will learn a lot from listening to (and talking to) your professors about those journals that are most important in the field. A recommended practice is to construct a list, relatively early in your Ph.D. career, of the top five journals in which you hope to publish. As the acceptances begin to arrive, cross titles off the list one at a time.

Once you have a general sense of where you would like to see your piece, you'll need to consider several practical matters before narrowing the field. For example, if the Shakespeare piece is over 6,000 words long, *SEL* won't publish it. Or, if you plan to be on the job market in five months, a particular journal's policy of taking six months (pretty much the norm these days) to respond to a submission simply won't work. Now there are two ways in which you can discover the sort of information that will allow you to make informed decisions. The first is to consult any one of the various "periodical guides" that provides information about journals in your field. These extraordinarily useful guides allow you to search information about journals relevant to your field or topic. The *MLA Guide to Periodicals*, for example, which lists information about most journals in the fields of English, Comparative Literature, and the Modern Languages, serves as a useful indicator of such guides' general effectiveness. A quick glance at "Shakespeare" in the index reveals that at least 11 journals and book presses include "Shakespeare" in their actual titles. If you then were to flip to any one of these journal entries, you'd be able to locate the following information: the editor's name and contact information; the history of the journal; subscription information ranging from the price of the journal to circulation numbers; advertising information; an editorial description of the sort of articles the journal publishes; and submission requirements. This final category is especially important for contributors, who can learn what the maximum length of submissions should be, the average amount of time before a publication decision is made, the average time between acceptance and publication, and revealing statistics about the easiness or difficulty of publishing a piece with the journal. For example, one would learn from looking at the *Shakespeare Quarterly* entry that although the journal receives 250 articles per year, it only publishes 16 of them.

Once you've chosen the journal you would like to submit an article to, the next step would be to consult that journal's own "For Contributors" page. More and more journals are providing an online version of this page. Even if you are completely satisfied by what you've found in a periodical guide, it's crucial that you consult the actual journal's guidelines. Journals often change their editorial policies and, more often, they change editors. You won't want to offend the new editor by addressing the old one in your cover letter. Once you determine that the information in the periodical guide matches up with the information in the journal itself, you should be ready to make a decision about where to send the piece.

Submitting the Article

The periodical guide and "For Contributors" page will also list the journal's preferences for style and mechanics. While no editor is likely to reject your piece because you've formatted according to *MLA* style when the journal uses *Chicago*, she might appreciate the fact that the piece appears as it would in the journal. I *recommend* formatting according to the journal's preferences; doing so should take no more than half a day (save the original formatting as well, since the piece may be rejected). If the journal practices anonymous submissions, make sure that your name appears nowhere in the article, including headers. Regardless of whether the journal finally prints footnotes or endnotes, manuscripts should always provide endnotes, double-spaced and beginning on a clean page. Now make the requested number of copies, including the electronic disk copy, and set the manuscript aside.

Your cover letter to the editor should be a very simple affair. Carrying on and on about your argument makes no sense for a variety of reasons but, most of all, because the cover letter is unlikely to get much further than a secretary's desk. Your job is to announce what you are submitting, to request publication in the journal, and to provide enough personal information that the editor will know how and where to reach you (make sure that you provide a professional e-mail address; again, hotpants@lovemail.com won't go over well). Figure 10.1 offers an example of a typical cover letter.

Always use your department's letterhead. You should list as your title, "Ph.D. Candidate." If you would like the manuscript to be returned to you, make sure you provide a SASE. I never bother to do so since I can't imagine sending a rejected article to another journal without first making changes to the manuscript. Once you've covered all these bases, your article is ready to go into the mail. You will probably receive an acknowledgment that your manuscript has been

---

UNIVERSITY LETTERHEAD

June 12, 2002

The Chaucer Review
Professors Susanna Fein and David Raybin
English Department
117 Burrowes Building
The Pennsylvania State University
University Park, PA 16802

Dear Professor Fein and Professor Raybin:

Please consider my manuscript—"Athletic and Discursive Competition in Fragment I of the *Canterbury Tales*"—for publication in *The Chaucer Review*. I have enclosed two copies as requested.
Should you need to contact me, I can be reached by phone at (860) 429-9106 or by e-mail at semenza@uconn.edu. My mailing address is listed below and on the first page of the manuscript. Thank you for your time and consideration.

Sincerely,

Gregory M. Colón Semenza
Assistant Professor of English

Figure 10.1 Sample cover letter for article submission

received within a few weeks after submitting it. If receipt has not been acknowledged after a month, call the journal to make sure the manuscript was not lost in the mail. Finally, you should keep in mind that you cannot submit an essay to more than one journal at a time. Some journals demand that authors include a line in their cover letters assuring the editor that their piece is not currently under review by another journal.

The Long Wait

Few things in academe are more irritating than awaiting an editor's response to a manuscript submission. After receiving the acknowledgment of receipt, you should prepare yourself for months of silence. Most journals still claim to return an answer within three months, but few are so timely in practice. Try to be patient. *The MLA Style Manual* advises that you may inquire about the article's status after four months. I usually wait for five months. If you decide to inquire, avoid allowing frustration or anger to pervade what should be a thoroughly professional correspondence. Whether by e-mail or snail mail, your inquiry should provide the title of your article, the original submission date, and your reason for writing, as demonstrated in figure 10.2.

Nine times out of ten, the editor will explain that the piece has been held up because (1) a reader is late in returning the piece or (2) the board has not yet met to discuss the reader's reports. An editor should respond in some way to your inquiries, however. If your e-mail or letter goes unanswered, try calling the office. I once dealt with an incompetent editor who, after holding my submission for more than a year, also refused to respond to any of my inquiries. Eventually, I withdrew the piece and placed it elsewhere, but only after having

---

UNIVERSITY LETTERHEAD

February 1, 2005

Dear Professor Jagger:

I am writing to inquire about the status of my article, "Sympathy for the Devil," which I submitted to *You Can't Always Get What You Want* journal on August 15, 2004. I would appreciate any information you might have regarding the piece.

Sincerely,

Lucifer

Figure 10.2 Sample inquiry letter to journal editor

wasted almost 2 years. Most editors will be as frustrated as you by delays since they make their journals look bad, and they typically will respond with as much useful information as they can reasonably provide. They also will appreciate your patience.

Now figuring four or five months as an average waiting period, you'll want to adjust your job market plan accordingly. Since your submissions may be rejected one or more times and since you wish to publish two articles as a graduate student, you should set the following goal: to submit *at least* two manuscripts *at least* one year prior to the time you plan to go on the market. Submitting them earlier would obviously be advisable since cutting things so close to the deadline will undoubtedly cause you a considerable amount of stress when you least need it.

The Decision Process

Journal practices and policies are various and sometimes highly idiosyncratic. Here's what you need to know. Once an editor receives your manuscript, she will glance at it to determine that it is appropriate for the journal and to decide which readers should and should not evaluate whether or not it should be published. If she decides that the piece doesn't meet the standards of the journal (or simply doesn't fit), she will send it back immediately. If the piece seems appropriate for the journal, she will send it on to readers, being careful not to select individuals who are either attacked or unduly flattered in the article. Here's a point that cover letters to editors can address; if your essay happens to be a critique of Marxist theory, you can ask the editor to please avoid sending it to readers who obviously will respond in a negative or closed-minded way. In any case, typical practice is to send a manuscript to two readers, though some journals consult as few as one and as many as five readers.

Readers are typically given a set of questions to guide their evaluations. Figure 10.3 offers an example of a typical form, in this case from the journal *The Eighteenth Century*.

Good readers will go far beyond a simple yes or no, taking time to highlight the strengths and weaknesses of any submission, suggesting how the piece might be improved, and recommending additional sources or relevant information. Generally speaking, the more quickly they read the piece, the more quickly you will hear back from a journal.

Once the editor receives all of the readers' reports, she must determine the next step. If both reports recommend that the article not be published, the editor will likely print out a rejection form letter and inform the author of the bad news. In most cases, rejections are

The Eighteenth Century:
Theory and Interpretation

MANUSCRIPT EVALUATION FORM

Title:

Reader:

Date:

Recommendation (check one):

A. Accept: ( ) Outstanding: ( ) Good: ( ) Acceptable
B. Revise (as specified below) and accept
C. Revise (as specified below) and resubmit
D. Reject

Please explain the reasons for your recommendations below or on an attached sheet; these will be returned to the author anonymously, unless otherwise requested.

Figure 10.3 Sample evaluator form (*for The Eighteenth Century: Theory and Interpretation*)

accompanied by the readers' reports or at least by a summary of those reports. Some journals refuse to provide authors any feedback, a practice that is obviously inconsiderate and unprofessional. If the readers' reports are split, the editor will send the piece to an additional reader or she will make a decision herself. If the editor determines that the piece should be published or finds that all readers' reports recommend publication, she still needs, in most cases, to bring her recommendation to the next meeting of the journal's general board of editors. Since boards do not meet all that regularly, they often are an additional cause of delay in the publishing process. Only after the board approves the editor's recommendation will she be able to inform the author of the journal's decision.

Reading the Response

Journal submissions are rarely accepted "as is" because good editors work hard to ensure that the strongest possible pieces are published in their journals; it almost always makes sense to request at least some revisions from the author. Generally speaking, editors will respond to your submissions in one of four ways:

Acceptance

Even if you are lucky enough to land an article "as is," you should make it a point to revise according to the reader's reports. While fortunately you will have the freedom to determine which advice to ignore and which to take seriously, regard the reports as indicators of how most readers will respond to your work. If both readers find fault with your handling of a particular issue, the smart thing to do is to address that issue. Your goal should be to publish the best possible version of the accepted piece.

Acceptance Pending Revision

A majority of your accepted articles will require some revisions. My general policy is to make any and all revisions recommended by the readers and editor unless such revisions alter the meaning or compromise the integrity of the article. Even if you determine that a particular revision suggestion is objectionable, you should try to meet the editor halfway; that is, if an acceptance depends on your revision of five passages and you oppose one of them, you should explain your objection and stress that you have been more than happy to address the other four points. As long as you are not being pigheaded, most editors will respect your decision. Always itemize your significant revisions so that the editor can more easily track and evaluate your changes. The appendix offers an example of an appropriate response to an editor's request for revisions, one which also will apply in relation to our next category (pp. 307–09).

Revise and Resubmit

Whereas an acceptance pending revision amounts to an agreement to publish your work so long as you revise it, an R&R suggests that the piece is not appropriate for publication *but* that the journal would be willing to consider a significantly revised version of it. Since editors gain nothing from encouraging resubmission of pieces they do not hope to publish, you can read an R&R as a positive sign of a journal's sincere interest in your work. As long as you determine that the revision suggestions are reasonable, you *should* revise and resubmit to the same journal as soon as possible. Often an editor will make a decision on the revised piece without even going back to readers. In other cases, she will send it to one or all of the original readers. In rare cases, she will send it to new readers. Chances are, though, that you will be more likely to receive a quick decision on a resubmitted piece than on an original submission. While acceptances are obviously preferable to R&Rs, you should remember that the only truly bad response from a journal is an outright rejection—and even that's fine as long as it's quick.

Rejection

Rejections can be helpful if editors are considerate. A few things to keep in mind: every scholar receives rejection letters. The quicker you develop the thick skin you will need to survive in academe, the more successful you will be. The key is in how immediately you are able to bounce back from despair. Upon receiving a rejection letter, consult the readers' reports and revise the piece accordingly. Once you complete the revisions and address the readers' concerns, send the manuscript to another journal. If you decide that the readers' reports are useless or unfair, ask a friend to look at the piece. What you think unfair, your friend may be able to explain in more palatable terms. If your friend agrees with your sense of the matter—and happens to be an honest person—then go ahead and reformat the piece for another journal and send it off.

Proofreading and Post-Acceptance Procedures

The first thing to do is have a drink. Experts have just recognized your work as being of the highest quality. Too often in academe, we move onto the next project without stopping along the way to assess and celebrate our achievements. My wife and I have dinner at a nice restaurant whenever another publication is added to the CV.

Once your piece has been accepted, you will likely be contacted by a managing editor whose responsibility is to oversee the proofing stage of the process. This stage will typically consist of three to four steps, depending on the journal's practices. Regardless of what form proofs take, you should make it a point to turn them back over as quickly as possible. Make copies at every stage so that comparison between documents is possible.

The Copy-edited Manuscript

Many publishers will send you a copy of the manuscript itself with editorial markings and marginal queries designed to guide your corrections and revisions. At this stage, authors are still permitted to make revisions so you should approach your review of a copy-edited manuscript as the last chance to make real changes; major rewriting would be inappropriate, of course, but two- or three-sentence length additions, deletions, and minor reorganizing all are acceptable.

"First" Proofs

Next, you will receive either galley proofs or page proofs or both (in two separate steps). Whereas galley proofs typically are translations of your text into single column printed pages, page proofs convert the text into printed pages that will eventually constitute the actual publication. Since substantial revisions are not welcome at this stage (changes are costly), you should understand that your job is to correct any mistakes that occurred during the transfer from your computer disk or manuscript to the publisher's printer.

Final Proofs

In some cases, you will receive one more set of proofs. Your job at this final stage is to make sure all mistakes have been corrected and that new ones have not been introduced in the process. You may also receive at this stage an order form for offprints of the article. Be sure to order at least twenty (I recommend 50 despite the cost) since you will want to send them to potential employers, colleagues who have assisted you in completing the article, tenure and promotion and merit committees, and influential people in your field. Some journals no longer provide offprints and others provide what basically amount to stapled photocopies of articles. You may wish to ask editors about their offprints if there's a possibility you'll save money by making your own high-quality copies of the article.

After returning the final set of proofs to the publisher, you will still have to wait several months before the article is actually published. All in all, you will find that the process of publishing an article takes at least a year and, in many cases, *several* years. Your hard work will all seem worth it, though, the first time you see your name in print.

Publishing Your Dissertation as a Book

The good news is that you will have accumulated experience working with publishers long before you need to begin worrying about publishing your first book. The surprising news is that you'll need to begin working on that book as soon as you are hired in a tenure-track position. Publishing a book—after it is written and completely revised—will take at least 2 years and probably more like 3 or 4, so you shouldn't wait more than 2 years or so to begin shopping it around. Selecting a publishable dissertation topic in the first place is, of course, the initial step in this process, as we discussed in chapter 8. Here I offer a few more tips about the process of turning your dissertation into a book, focusing especially on the creation of a prospectus.

Envisioning the Book

Routledge editor and author of *Getting it Published*, William Germano, rightly emphasizes the differences between dissertations and scholarly monographs, claiming that most dissertations simply don't make good books. The ones that do tend not to look much like dissertations in the first place: "What an editor is looking for—and sometimes does find—is the book you happened to be writing as you were writing your dissertation."[3] He cites as an example of the problem the long "Review of Literature" chapters that introduce most dissertations (see pp. 182–83). He also complains about the "thesis-plus-four-applications" format employed by many dissertators. That is, the format by which one offers a topic in the opening and then explores it in relation to four texts or case studies in subsequent chapters.

While I agree wholeheartedly with Germano's sense of the difficulties facing revisers of dissertations, I am convinced that the problems are institutional and psychological as well as formal. Over the past 20 years, a major shift has occurred in the academy from a situation in which professors sometimes *chose* to publish their dissertations—often many years after tenure—to one in which young faculty members are *expected* to publish their dissertations immediately. Even though the new model has been in place for more than 20 years, many major advisors continue to treat the dissertation as an animal only loosely related to the monograph. Not all Ph.D. students are ready to write books, it's true, but they should at least be encouraged to write something that looks like a book. What I am suggesting, in other words, is that the old dissertation—with its massive review of literature and "thesis-plus-four-applications" model is extinct. The dissertation has evolved. And in the new academic world, the old dissertation is simply unfit for survival.

One institutional solution to your future problems, then, would be to choose your advisors very wisely (see pp. 45–47). Ask potential advisors how they understand the relationship between the dissertation and scholarly monographs. Ask them whether or not they feel comfortable (and capable) enough to advise a *book* to completion. Ask them what they believe constitutes a reasonable plan for publishing that book once you are hired on the tenure track. From their answers you will be able to extrapolate much about their likely effectiveness as major advisors, and you will be able to make an informed decision accordingly.

The anxieties of many graduate students also explain a lot. On a certain level, the excessive "review of literature" serves as an obvious attempt on the student's part to say, "here's what I know. Look at all of these books that I've read." The "thesis-plus-four-applications" model stands in for an actual argument, which the student may lack confidence enough to pursue. Such anxieties and failures of confidence are likely exacerbated by advisors who draw attention to the differences between books and dissertations: students know they *should be* writing a book but are coached instead to write an archaic, impractical document, which reinforces their sense of inadequacy regarding the ability to publish a monograph.

Most academics continue to experience "impostor syndrome" long after graduate school. You're not alone, in other words, if you feel anxious or uncertain about your readiness for writing a book. Remember, though, that I am recommending not that you seek to produce an immediately publishable dissertation but, rather, that you write a dissertation that looks and feels like a book. You still will have years to revise it into adequate shape. Merely by envisioning your project as a book, though, you will likely cut down on the sorts of problems that Germano and other editors find so problematic about dissertations. Seize the energy and excitement that comes from imagining your first monograph and channel it into the daily work of writing. By choosing a savvy major advisor willing to guide your writing, and by contending directly with your own anxieties, you'll be more likely to fashion a voice that is your own and a document that can eventually translate into a monograph. (For more information about how to create such a document, see chapter 8.)

Revising and Expanding

Your defense hopefully will alert you to the revisions you'll need to make before your dissertation can become a book. Often dissertators discover provocative new material for additional chapters while conducting their original research; expansion is therefore one of the most common forms of revision. Other dissertators simply need to strengthen the connections between the chapters or to add more contextual information. Regardless of what form your revisions take, don't wait long to begin revising. The summer before beginning your new job is the best time to start, regardless of how distracted you may be by the upcoming move. Even if you are only able to work up a revision schedule and a long-term plan for publishing the book, your time that summer will be well spent. You'll find it extremely difficult to get much new work done the first year or two on the job. Stress levels alone will be enough to affect quite negatively your usual levels of productivity. Revisions are quite possible, though, since they usually involve improving what you've already created.

How Much to Publish

Above I recommended that you publish at least one chapter of your dissertation prior to going on the market, which implies that more is better. But how much is too much when it comes to the publication of chapters as articles? Young scholars face something of a unique predicament when it comes to answering this question. Since book publishers are less willing to take chances on unproven writers, they will like to see that your work has been given the stamp of approval, so to speak, by top journals. No book publisher wants simply to reprint already published material, though, and so most presses refuse to publish books from which more than 25 percent to 30 percent of the material has been previously published. Therefore, you should aim to publish no more than two chapters from your dissertation.

Selecting a Press

Just as you learned which journals to aim for, you'll begin to learn over time which book presses are most appropriate for your work. Obviously there are several major presses for humanities scholars—Oxford, Cambridge, Chicago, and so on—that are likely to attract your attention regardless of what you happened to write your dissertation on. But beyond the behemoths, the factors that influence one's decision of where to publish can be highly idiosyncratic and field- or topic-specific. Whereas Bucknell University Press, for example, has little prestige for medieval scholars, eighteenth-century literary scholars rightly view it as a strong outlet. In the old days, accepted wisdom regarded university presses as superior to trade presses. The ascendancy in recent years of several powerful and highly competitive trade presses—for example, Palgrave Macmillan, Routledge, and Ashgate—has significantly altered such perceptions. In fact, trade presses are often far more efficient than university presses, and they tend to have superior resources for advertising and marketing. As in the case of trying to select journals, you'll need to pay attention to where your colleagues are publishing and talk with your advisors about those presses that are right for your work.

Construct a list of 15 or 20 presses with which you'd consider publishing your work, and try to arrange them in order of preference. Once you have created your prospectus, you can begin sending it to four or five of these presses at a time. Avoid sending 20 prospectuses at once for two reasons: (1) you may receive early, useful feedback from editors than you can use to revise the prospectus; (2) you may attract the attention of more than one editor. Let's say that Cambridge and Illinois both agree to read your book: because presses typically demand the exclusive right to evaluate a manuscript, you thank the editor at Illinois for her interest and decide to submit the book to Cambridge, which rejects it seven months later. Now you can't go back to Illinois.

You may also find your choices narrowing over the years for largely unpredictable reasons. Perhaps you will meet an editor at a conference who will express interest in your work. Or maybe a particular press will begin to specialize in books directly related to your dissertation topic. In my own case, an opportunity emerged shortly after I was hired which involved a strong university press's willingness to read my manuscript. Having not yet mailed a prospectus to Cambridge, Oxford, or Chicago, I was faced with a dilemma: should I delay responding and thereby risk turning off the interested press, or should I recognize the difficulty of publishing a book—especially for first-time authors—and go ahead and submit the manuscript? I decided to submit the manuscript. I have never regretted my decision, especially since it allowed me to publish the book early on in my professorial career, but I do not think the other decision would have been wrong either. The point is that you should construct a plan, but you should also recognize the likelihood that you will veer away from it for one reason or another.

Preparing the Prospectus

Once you develop a plan for approaching presses, it's time to begin writing the prospectus. A key feature of any persuasive prospectus is the author's ability to convey the project's contribution in a concise and clear manner. Prospectuses, therefore, should be about five pages excluding the cover letter, though some presses will allot authors as many as 2000 words, which is more like seven pages. Remember that while your extremely focused, perhaps even cutting-edge project might go over really well with colleagues at a conference, editors will be focused on numerous practical considerations. How much need is there for such a book? Is this book likely to sell? How much work will the press need to do to ready this book for publication? Whereas jargon may sound intelligent to you, it will likely suggest to an editor your inability to communicate clearly. Whereas you may view the noncanonical subject matter as progressive, an editor might see it as unlikely to sell. In constructing a prospectus, you must communicate the marketability of the project without surrendering its intellectual integrity. In what follows, I've broken down into typical components an academic book prospectus, two of which can be found in the appendix (pp. 312–18). I do not wish to suggest that all prospectuses must include all of these parts, and the order in which prospectuses arrange the parts also tends to vary. You will find that most presses have clearly defined policies about what they wish to receive from prospective authors. Always consult a press's web site before contacting an editor.

Cover Letter

Your one to two-page cover letter should be simple and to the point: "I have written a book on such and such a topic. Would you be interested in publishing it?" If you have had previous contact with the editor, make sure you remind her of this fact in the opening paragraph. The rhetorical style of a cover letter will vary from author to author, but as figure 10.4 reveals, a solid cover letter will always convey certain important pieces of

---

January 3, 2002

Ms. Kathryn Amanarindo
Assistant Acquisitions Editor, Humanities
The Greatest University Press
2715 Charles Street
Bethesda, Maryland 21218–4323

Dear Ms. Amanarindo:

I am writing to inquire whether you would be interested in reviewing my book manuscript, "Sport, Politics, and Literature in Early Modern England," for publication by Greatest University Press. My interdisciplinary work on Early Modern culture and literature ties in nicely with recent GUP monographs on the cultural history of the period, including Ralph Falcon's *Charismatic Criminals*, Paul R. Backson's *Spectacular Sites*, and Tonya Sere's, *Village People*.

Despite recent critical interest in nearly every aspect of Early Modern English popular culture, scholars have ignored sport, exercise, and athletics. This neglect is puzzling since sport occupied an integral position—both literal and metaphorical—in politics, medicine, military science, and art. To the degree that Early Modern scholars have studied "sport" at all, they have tended to conflate athletics and mirthful, disorderly activities such as drinking and gambling. In contrast, my book demonstrates that sport was central to Early Modern conceptions of order, health, and nobility, and it shows how major writers like Shakespeare and Milton used contemporary controversies about sport as a vehicle for social commentary and protest.

The critical response to this project—from colleagues in English, History, and Comparative Literature—has been enthusiastic. Several preliminary ideas are developed in essays published, or accepted for publication, in *SEL*, *Renaissance Quarterly*, and *Prose Studies*.

The completed manuscript, including the bibliography, is approximately 320-pages typescript (78,076 words, excluding bibliography) and requires no special design attention.

I hope that you will be interested in reviewing my book manuscript. Enclosed you will find a brief prospectus, chapter outline, introduction, sample chapter, and vita. I can be reached by phone at my office (861 486-4723) or home (861 429-9096) and by e-mail (semenza@uconn.edu). Thank you for your time and consideration. I look forward to hearing from you.

Yours Sincerely,

Gregory M. Colón Semenza
Assistant Professor of English

Figure 10.4 Sample cover letter for book prospectus

information: first, the title and subject of the book (see also pp. 309–12). What is the book about? What sort of audience is it aimed at? Second, it will describe the current status of the manuscript. If the book is not complete, let the editor know when you plan to finish it (for a first book, I recommend sending prospectuses only after you have completed the manuscript). Finally, be sure to state your credentials for authoring the manuscript.

The letterhead will convey your affiliation, but you should make clear your rank as well. A quick mentioning of relevant publications will further demonstrate your qualifications; especially if you have already published parts of the proposed book, say so, since such publications speak to the quality of your project.

Project Description

Avoiding jargon, explain what your book is about. Be sure to highlight your central argument since the editor will want to know how your work advances or contributes something to an existing conversation. The good news is that the job market will have prepared you for summarizing your work in a concise and clear manner. You may even find yourself able to cut and paste sections of your job application or dissertation abstract. Just remember that you should remove all remnants of the "dissertation" from an actual book prospectus. For an example of a project description, see the Appendix.

Audience

In a few paragraphs, explain who will want to read your book. Obviously a larger audience will be more appealing to a press. Especially if you envision the book being useful to teachers or students, say so, since regular classroom use of your book will translate into regular sales for the press. Be very direct in this section since you need to make the case to a press that the book will sell. Consult the prospectuses in the appendix, in addition to studying the example in figure 10.5.

Competing Books

Are there published books similar to your own? If so, in what ways is your book different? In a paragraph, explain. See figure 10.6. In many ways, the "Competing Books" section demands information you should already be quite able to articulate: how is your book to be situated within a larger discourse community?

Chapter Summary or Table of Contents

Should you decide to provide more than a table of contents, limit your description of each chapter to no more than a short paragraph,

There is also a much wider audience for this book, consisting of historians, Dickens enthusiasts, and those interested in imprisonment and the origins of modern psychology. In many ways, our culture has not outgrown the Victorians' fascination with penitentiaries and criminals—with confessions, secret horrors, prison scandals, and the private infliction of insanity. Recent movies like *Murder in the First*, *Dead Man Walking*, and *The Green Mile* bear that out, as do new books by Norvald Morris and Peter Brooks. Though *The Self in the Cell* is "about" a serious scholarly topic, then, it has the power to attract thoughtful non-scholars as well. Just as important, the manuscript is written throughout in language that is sophisticated but accessible to the array of audiences who may take up the book. Readers of *The Self in the Cell* will make important discoveries about Dickens, Charlotte Brontë, and other Victorian novelists, to be sure. But they will also, I believe, come to fuller appreciations of the prison's importance to the development of interior narrative, psychoanalytic practice, and the shape of the modern novel.

Figure 10.5 Sample prospectus audience description

Although *Milton in Popular Culture* would have no direct competition, it can be most closely compared to books on Shakespeare and popular culture or to Milton companions and teaching volumes. The most relevant previous publication is Douglas Lanier's *Shakespeare and Modern Popular Culture* (Oxford, 2002). Along with such groundbreaking books as *Shakespeare, The Movie* (Routledge, 1997), *Shakespeare and the Moving Image* (Cambridge, 1994), and *Shakespeare After Mass Media* (Palgrave, 2002), Lanier's work demonstrates the sort of broad interest a study of a canonical figure and popular culture can generate. Unfortunately for Renaissance scholars, such studies have been limited only to one author: Shakespeare. Studies of Milton most comparable to this volume would be Thomas Corns, ed., *A Companion to Milton* (Blackwell, 2001), Richard Bradford's *The Complete Critical Guide to Milton* (Routledge 2001), and Peter Herman's (ed.) forthcoming *MLA Approaches to Teaching Milton*, all of which are aimed at a broad academic audience but none of which deal with popular culture.

Figure 10.6 Sample prospectus "competing books" paragraph

similar to those represented in figure 10.7. As an alternative to creating a separate chapter breakdown, you may wish to describe the contents of the book in the project description itself.

In addition to the cover letter, many presses will request a sample chapter or an introduction. In considering which chapter to send, think both about which chapter is the strongest and which is likely to appeal to the widest audience. Never send an entire manuscript to an editor. The editor will never read it, and she may be annoyed by your ignorance. Once the document is out of your hands, try to exercise the same patience you've learned from dealing with journals.

Chapter Three: "The Literary Context of the *Book of Sports* Controversy" demonstrates the mutually constitutive relationship between literary and political commentaries on sport. Investigating closely several anti-court satires written in the 1610s—including *Eastward Ho* and *The Isle of Gulls*—the chapter elucidates the manner in which dramatists used sport to critique or defend the political policies of James I. The primary focus of the chapter, however, is on James' *Book of Sports* as a reaction to such dramatic commentaries. In short, the king's defense of lawful pastimes is a deliberate attempt to counter his popular reputation as an unlawful monarch.

Figure 10.7 Sample prospectus chapter description

CONCLUSION

The proverbial pressure to publish can indeed be paralyzing, but it need not be. Like the development of most skills, learning to publish requires knowledge, practice, and multiple failures. The failures and the rejections, you should know, are unlikely ever to cease completely. But such challenges are precisely what make professional writing so stimulating an activity and publishing so rewarding an accomplishment.

Chapter 11

Service and Participation

Of the three most important activities performed by the majority of graduate students—service, teaching, and research—service is the least important. In the 2001 scientific survey of "What Search Committees Want," service was ranked 3.42 on a scale of one to six—slightly less than important, in other words.[1] This fact is quite misleading, however, since service will be of major importance at the assistant professor level, especially for those individuals hired by smaller colleges and universities. Paul Hanstedt claims that in many so-called teaching colleges, "service almost surpasses teaching effectiveness as the main means of establishing institutional suitability."[2] Since "Potential for making a positive contribution to the institution as a whole" receives the highest ranking of any category (5.36), the study sends the clear message that although your service record won't get you hired, it will make you a considerably stronger job candidate for all types of institution.[3] How you approach your service obligations as a graduate student not only suggests to potential employers how likely you are to contribute to their departments should they hire you, but also how efficiently and effectively you will be able to do so. In this relatively brief chapter, I consider the following issues related to institutional service and participation:

- Differentiating useful and useless service activities
- Participating in specialized-field activities
- Maintaining efficiency in committee work
- Avoiding "service exploitation" as a minority student
- Taking on student leadership roles
- Understanding service obligations after the Ph.D.

While leadership and diplomacy skills are probably not easy to acquire later in life, an awareness of typical service demands and a practical plan for dealing with them can help you to fashion yourself as an invaluable contributor to your college or university.

TOWARD A PHILOSOPHY OF ACADEMIC SERVICE

"Service" is perhaps not the best term for describing one of the most important activities performed by most university professors; unfortunately, it connotes mainly the act of "serv*ing*" someone else or, worse, serving the rather abstract entity called the "university." If we were living in the Renaissance, when coining one's own terms was more common, I'd start a movement to replace the negative word "service" with a word like "participation," since the latter has the significant advantage, at least, of granting participators a certain degree of agency. Slaves *serve*. Servants definitely *serve*. Individuals participate. More important, they benefit considerably from their participation in decisions that directly affect their own personal and professional lives. "Participation" is a superior word for the additional reason that common activities we consider under the "service" umbrella don't always happen to take the form of committee work; your active attendance at a department meeting, your presence at a guest lecture, and your willingness to pick up an extra cup of coffee for the administrative assistant all constitute forms of positive participation in the culture of your department.

If there are terminological problems with the word "service," there also are categorical and conceptual ones: "service" is used typically to distinguish certain activities not directly related to research and teaching. The implication is that service functions in a compensatory manner, forcing professors to pay their dues so that they might go on enjoying their weird little obsessions. One might argue that since most academics have deliberately chosen to forego the road more traveled (that is, the road to personal wealth) in order to become educators, almost everything they do is in the *service* of the greater good. Rather than differentiating a professor's excellent teaching and the curricular decisions that help him to be an excellent teacher, we should approach both as contributing to the same basic objective, which is to educate our students in the most effective ways we can. If we were to promote overall "good citizenship" rather than advocating disproportionate competence in the separate areas of service, teaching, and research, our universities might be more equitable and satisfying places to work.

Of course, not all professors care that much about equity; a small percentage of them have managed to publish a great deal, earn an awkward sort of "fame" in their own subfields, and make a lot of money precisely by refusing to perform those activities that don't benefit them personally. For some, such a strategy is apparently something to be proud of, as suggested by College of New Jersey professor, David Lester's recent testimony in *The Chronicle of Higher Education*:

> I have made some decisions over the course of my career that have allowed me to be productive, yet not feel overwhelmed. I went to the first graduation ceremony at the college in 1973, but I have never attended one since. I have not attended a faculty meeting since 1972. I found that I liked my colleagues much better if I did not listen to their silly comments in such meetings. I rarely go to division meetings . . . but I do try to make most meetings of the psychology program. . . . For many years, I had my name removed from the faculty e-mail list so that I had no awareness of what activities were taking place at the college. . . . I do not pick up the telephone in my office, and my voice-mail message informs callers that I do not check for telephone messages. . . . I have avoided as much college service as I can. . . . Our pay raises . . . do not depend upon evaluations by a dean or other administrator.

If we were to summarize the piece, Professor Lester's statement might look something like the following: "By neglecting my departmental and college duties and refusing to assist in the governance of the institution largely responsible for my own successes, I've managed to publish a lot." I suppose that one can only feel bad for Lester's colleagues, students, and especially his department head. Unfortunately, the consequences of his behavior may be significant for other departments and colleges as well; because the Department of Psychology at the College of New Jersey tolerates Lester's behavior, some graduate students in his department might believe that acting selfishly is generally acceptable, and they might even try to emulate his behavior once they are hired at the assistant professor level. For the vast majority of them, this would be a terrible mistake.

My advice is that you approach "service" as an extension of your teaching and research activities—not as a separate or less important activity. Although the current system makes it easy and even somewhat necessary to separate out and hierarchize teaching, service, and research, remember that, first as a job candidate and later as a pretenured professor, you will be judged according to your ability to contribute positively to your department and/or university. There are very few exceptions to this rule. Service work that you refuse or simply fail to perform will translate into extra work for others in your department. Such negligence will provoke justifiable feelings of resentment from the people who actually do contribute, and it will almost certainly affect your tenure case. If you can't see the ethical reasons for contributing, try at least to see the practical ones.

What you will very likely find is that good citizenship brings with it many rewards, not the least important of which is a respectable departmental voice and a more formidable ethos. Whether as a graduate student or professor, your work on various committees will give you the right to speak on matters important to you and your colleagues. The information you accumulate from working carefully on a particular issue will help to instill confidence in you, which will allow you slowly to begin plugging all those gaps in your knowledge of the department's history and politics. Such knowledge will empower you, in turn, to speak out on the issues that trouble you most and offer suggestions when good ideas strike you. The harder you work to affect changes that address your personal and professional needs, the more rapidly your institution will come to resemble your ideal workplace. You may be surprised by the satisfaction that comes from teaching and studying in a system you have in some small way helped to create.

THE POLITICS OF COMMITTEE WORK

Committees tend to be tricky animals because, in most cases, the members that compose them are so utterly various, each one owning a different level of power and experience. On graduate-student committees, for instance, you may find yourself working at once with ABDs and first-year MAs, persons with extremely different research and teaching interests, and even persons from other disciplines, including those in the sciences, engineering, and business. Departmental committees involving both faculty and graduate students could have you working in a service capacity alongside your department head or even your dissertation advisor. If you were to serve as a student representative on a college-level committee, you might find yourself surrounded at a table by the dean, the associate dean, and several important faculty members from various colleges. The point is that such diversity represents both what's wonderful and what's terrifying about committee work. On the one hand there's a great deal for you to learn and many experienced people for you to learn from; on the other hand, trying to determine your role and responsibilities in such situations can be extraordinarily difficult and even a bit overwhelming. In approaching your committee responsibilities, keep in mind the following advice, and remember that learning in graduate school how to negotiate the complex politics of committee work will give you serious advantages on the job market and especially on the tenure track:

Listen and Learn: Learn about how the specific committee operates before going too far out on a limb. From one meeting and a few discussions with experienced members of a particular committee, you'll learn just about all you need to know about how the committee works. Sit back and observe those committee members that seem to be the most informed about the issues and most widely respected by others; take note of the degree of formality with which the meeting is run; pay attention to official procedures for conducting business and, especially, to the necessary steps that must be taken for voting in any new policies. A very good idea might be to volunteer to take minutes at the committee's first meeting. By doing so, you'll simultaneously position yourself to observe the manner in which the committee is run and make clear your willingness to participate actively in it. Often well-intentioned, new members simply rush into the committee room, eager to change things for the better, but usually before they really understand the complexities of the specific situation. Relax and know that there will be adequate time ahead for making your presence felt.

Do Your Part, and Above All Else, Do No Harm: I observe regularly as a teacher that whereas excellent students despise "group projects," lazy students absolutely love them. Obviously, the best students understand that, in order to maintain their superior standards of excellence, they will need to work even harder than usual to compensate for the incompetence of their less ambitious classmates. Like my best students, I am constantly irritated by the unfairness of the group dynamic, though I continue to impose it on them semester after semester. I realize, after all, that few lessons prepare them better for professional life than having to confront rather early on the unfortunate fact that the real work in most institutions happens to be performed by very few individuals.

As Professor Lester's celebration of negligence (see page 240) would suggest, academe is no more exempt from such frustrating realities than any other professional realm. You should remember, therefore, that any committee work you fail to do will need to be done by your colleagues. Remember also that colleagues will very likely hate your guts for having to do your work. Now think about how dangerous and impractical it is to be disliked by anyone in a world as self-contained and political as academe. Use your common sense. If you've just received forty papers and feel overextended, ask the committee chair for relief from a particular assignment, explaining that you will make it up the following week. If you know that you don't handle certain types of responsibilities particularly well, volunteer early in the semester to work on other things. Since simply not doing the work is not an option, you should be strategic about how you get the work done.

Of course, Negligence is not the only opposite of participation. The most important advice I can give you is to avoid developing a reputation as a "difficult" person. Always remember that the single most important thing your advisors must be able to say about you when you enter the job market is that you will contribute *positively* to the life of a particular academic institution. While constructive criticism should always be welcome in academe, and while reform should always be viewed as an admirable goal, some well-intentioned graduate students damage their reputations by generating divisiveness where they should be promoting unity, by causing anger where they should be stimulating thoughtfulness. I once witnessed a well-respected Ph.D. candidate damage in about 30 seconds a reputation he had taken years to build, by launching highly personal attacks against students and faculty in an e-mail to the entire department. The reason for his incivility—which led him to press "send" when he should have pressed "delete"—was an easily resolvable difference of opinion about a new and rather trivial committee issue. Feelings were immediately hurt, several nasty exchanges took place both in private and in public spaces, and within days, the department was divided over an e-mail that should never have been sent. I could only sit back, embarrassed for everyone involved, and imagine how difficult it was going to be for that student's advisors to write straight-faced letters of recommendation.

Avoid Being Unnecessarily Deferential: While it would be unwise to forget the power relationships in play on any committee—especially one including faculty members and administrators, as well as graduate students—you also should keep in mind that you are an important member of the committee. Especially if you are serving in the capacity of student "representative," it would be remiss of you to remain silent as others conduct business very likely to affect your constituents. Within the walls of the committee room, try to see yourself as a colleague, a member of a group working on shared problems. By participating actively and intelligently in the affairs of the committee, you'll earn the respect of those colleagues and influence the work that they do.

Avoid Trash-Talking: One of the dangers of representing your department in college- or university-level committees is that your "loyalties" can so easily become divided. Try to remember, though, that at least as long as you are a graduate student and assistant professor, your primary loyalties must be to your department. Its inhabitants will educate you, employ you, and, most important, recommend you for work and/or tenure. Often it will be tempting to share with your new pals outside of the department news of recent departmental gossip and scandal. Be careful. Word travels very fast in campus circles, and you wouldn't want your advisor or department head to find out that you are the reason why everyone in the Physics Department knows about Professor Lewd's recent affair with one of his freshmen. Remember that the relationships you build in your department will remain important to you long after you forget the names of those colleagues with whom you happened to serve on the university's "library circulation policies committee."

Build Strong Relationships (i.e., "Network"): Or maybe you'll seek to cultivate more lasting relationships with such colleagues. While you'll always want to be careful about how you represent your department and its members, there's certainly no harm in pursuing strong extra-departmental relationships, which can be professionally useful as well as personally rewarding. You should approach the chance to work closely with a dean, for example, not as an opportunity for screwing up and sinking your career, but rather as an opportunity to impress a rather important person in your university. Excellent service will make you and your department look good, which can translate positively for you both inside and outside of the department.

SELECTIVE SERVICE

All committees are not created equal and, as a graduate student, you have considerable leeway to decide which ones you'd prefer to join. On the one hand, graduate student service is absolutely voluntary, in that no one can force you to do it. On the other, failure to perform at least some service activities as a graduate student can be disastrous, since potential employers will expect to find some evidence of life under the "Service" heading of your CV. The key is to recognize that because you have freedom of choice in this matter, you *can* participate only in those activities that benefit you professionally and personally. So what are those activities? While they will vary from person to person, you might consider the following ideas while seeking to build an impressive service portfolio.

Be a Force in Your Field

Above all else, your goal should be to contribute to the community of scholars working in your field of specialization. Only Venn diagrams could effectively represent the numerous, complex relationships you'll develop over the course of your graduate career. You'll need to cultivate during this time relationships with undergraduates, other graduate students, staff members, and faculty and administrators both within and outside of your own department. In the end, though, the most important relationships will be with your dissertation advisor and associate advisory committee members. Far more crucial than any service category which you might be able to check off on a CV will be your advisors' testimonies about your willingness and effective ability to contribute to the culture of the department. Without actual proof of such willingness and ability, your advisors' hands will be tied; they might be able to defend the integrity of your research or comment on the quality of your teaching, but their letters will not be able to address very successfully what potential employers will most want to know. The difference between a faculty member saying "I believe that Ms. Johnson will be able to contribute" and "Ms. Johnson has contributed much," may also be the difference between a tenure-track job and another year without one.

Here's what you can do. Show up at any on-campus talks given by visiting experts in your field. Attend any meetings held to discuss field-related matters and any relevant social events to which you may be invited. Participate in and consider organizing reading groups, on-campus conferences, and departmental colloquia related to your subject of study. Ask your professors to involve you in job searches for candidates in your specialized field. Help out in recruiting graduate students likely to specialize in your area. In these ways and others, you can demonstrate to your advisors how invested you are in the good of the larger community, making their jobs much easier, in turn.

Graduate Student Government

It's safe to say that graduate student representatives are typically regarded with some degree of ambivalence. The common suspicion seems to be that each year's election amounts to a popularity contest similar to the process of selecting the prom king and queen. While there may be something to such a claim, my own suspicion is that most of the ambivalence happens to be felt and articulated by people who really are reacting to something else—whether their own unwillingness to participate or their envy of those who have a say in the daily governance of their world. The simple fact is that most competent persons who *really, really* want to be elected can figure out a way to be elected. Some very mature individuals recognize that one reason to serve is in order to reshape what they perceive to be the organization's flaws so that it can more effectively address graduate students' needs.

While members of graduate student organizations range from slavish defenders of the status quo to dedicated reformers and representatives, they all strengthen their professional qualifications by choosing to serve. Not only do they show a willingness to *volunteer* their time and labor, but they also gain invaluable experience in negotiating the politics of the academic committee. Further, they may actually play an important role in the daily governance of their departments and universities. The president of the representative student organization in my department, for example, serves on the graduate executive committee (responsible for every important decision affecting graduate students, including course scheduling, curricular reform, and admissions), the graduate examination committee, and various search committees. Such a person will enter his first job with an important, inside-view of how departments actually are run. Obviously, this means he has considerable advantages on the market.

Administrative and Professional Posts

Opportunities often arise in which graduate students are able to trade their teaching appointment for administrative or professional posts. For example, a student might serve for one or more years as an academic advisor for undergraduate majors in the department. During this time, the student either would not teach or would teach less. The advantages of some such opportunities are significant; in this particular example, a student would enter the job market having advised dozens—in some cases, hundreds—of students, experience that most universities will value highly. The only problems with such posts tend to emerge when students are allowed to keep them for longer than a few semesters. In such cases, the department benefits from employing what basically amounts to a full-time staff member at about a third of the cost. Unless the student plans to become an administrator, the post ceases to help him after a few semesters, since teaching experience is far more important to most employers. If you should have an opportunity to diversify your work experience, do not hesitate to do so, but limit the time you are away from teaching.

Particularly useful opportunities, other than advising, include any work related to professional societies, conferences, and/or journals. Working for a professor who edits a major journal, especially one in your field, allows you to see from the inside how journals operate and increases your understanding of what types of essays get published and why. Helping a professor to organize a conference will likely result in a wider network of professional acquaintances, some of whom may be influential scholars in your field. Finally, many departments regularly appoint RAs or "Research Assistants." Usually in exchange for your regular teaching stipend and tuition remission, if appointed an RA your job will be to assist a professor in researching material for something like a new book project or a new course. Talk to the professor about what will be expected of you; if your job will be merely to fetch books at the library, you might want to pass. If assisting the professor will help you to learn something—either about the material you will research or activities like book publishing and course development—you should consider taking the job. Just be sure to return to teaching after a semester or two.

Community Service

When I asked a college dean (somewhat naively) about community service during one of my campus interviews, he smiled kindly and replied: "Community service is very nice, but it won't get you very far here." Fair enough. I suppose one doesn't perform community service for purposes of professional advancement. At the same time, my question was intended to gauge how this particular university viewed its relationship to the nonacademic community, to discover whether it placed any value whatsoever on outreach. While the answer clearly was "no" in this particular case of a large research university, the simple fact is that many colleges and universities do value outreach immensely. Administrators at smaller colleges especially may find quite appealing a history professor willing to run a reading group at the local bookstore or an art history professor willing to guide museum tours for local kids a few times a year. Such activities enhance the reputation of the institution amongst local citizens and may even assist in strengthening ties with local business owners. For the most focused and ethical administrators, such activities simply deserve to be rewarded because they live up to the ideals of the liberal arts mission—which is to educate. Hopefully you'll continue to participate in your community whether your university supports such service or not; I bring up the issue here simply because you may wish to pursue work in an institution that places a higher value on such service.

Mandatory Service

Once you are a faculty member, service will no longer be optional. Nor should it be a burden if you approach it sensibly and systematically. Many, if not most, departments *claim* that their first-year assistant professors are spared the burden of committee work while they adjust to their new surroundings. In practice, though, most departments benefit immensely from the service of their first-year professors, who usually are energetic and quite willing to involve themselves in the life of their new workplace. Often what separates the most obvious tenure candidates from the most controversial ones is the sense conveyed by the former that they are invested in the greater good of the department and the university. Occasionally misguided assistant professors enter their new jobs with an overinflated sense of their own importance; after months of having been courted on the job market—a time period during which their ideas and work have been constantly at the center of attention—they lose sight of the fact that they must start anew in trying to earn the respect of their colleagues, students, and superiors. I have witnessed several situations in which immature beginning assistant professors have managed to alienate their colleagues by refusing to take their service responsibilities seriously. Such persons provoke the animosity of those people that matter the most in their professional lives: that is, the people who will vote on their tenure cases a few years later.

There is a positive side to all of this, of course: by demonstrating your genuine commitment to service, you can suggest to your colleagues that you understand your role as a member of a community of scholars and teachers, a significant first step toward building a strong reputation in your department. By accumulating service experience as a graduate student, you'll also have a head start in learning to deal with the complex issues that tend to arise in academic committee rooms. In fact, if your experience is significant enough, you will encounter very few surprises as an assistant professor.

In what follows, I discuss rather briefly some practical guidelines for approaching committee work as a new assistant professor. One could argue that such guidelines might be more appropriately addressed in a book written for assistant professors, not one for graduate students; I've decided to touch on these issues, nonetheless, for two practical reasons: first, because you'll need to answer multiple service-related questions during job interviews; second, because paying attention now to the sorts of issues you'll confront on the tenure track increases the likelihood that you will be adequately informed when it comes time to make more regular contributions.

Keep a Service Log

Due to the sorts of service abuses that oftentimes occur in academe, you'd be wise to keep a service log once you start your new job. In it you should track not only the hours you spend in committee rooms, but also any time you spend on service-related work in your office or in any other space. Include in the file e-mails and other documents that support your records.

Stick at First With What You Know

During my first semester as an assistant professor, I volunteered to serve on the job placement committee for graduate students. I realized that my recent experience on the market made me a potentially useful resource for them, and I saw the committee as a way to demonstrate simultaneously my commitment to graduate students and my desire to participate in the life of the department. Had I served that semester on a different committee, it would have taken me much longer to feel settled in my new environment; instead, I learned a great deal about the graduate program and its students' needs, gained opportunities for meeting a relatively large number of faculty members and students, and gained confidence that I could contribute something valuable to my new department. My advice: if hired by an MA- or Ph.D.-granting institution, involve yourself right away in the graduate program since this is an area about which you know a great deal.

Demonstrate Interest

In a way, this is like saying, "Do the bare minimum." Since not all people succeed in doing this, however, it's worth a little space here. Most important, be present at department meetings. Even if you choose not to participate much at first, your regular presence will be noted, and you will learn much from what goes on. If your department hosts beginning-of-the-year or holiday parties, make it a point to attend. If a particular group of faculty members gets together for martinis once a month, join them a few months out of the year. Agree to attend undergraduate commencement every few semesters. And obviously apply the same rules about activities in your field of specialization as you would as a graduate student: attend talks, participate in reading groups, and so on. In other words, show yourself at all times to be collegial and engaged.

Choose Useful Committees

I don't mean to imply by "useful" that most departments have "useless" committees. Rather, I wish to suggest that certain committees are ideal for new faculty members because they represent opportunities to learn a great deal about the department in a relatively brief period of time. If given the chance, you'll benefit from serving on any of the following committees:

Department and/or Graduate Executive Committees: Since the most important decisions usually are made by the executive committees, you'll gain an inside-view of how things work in your department by serving on them. Because the executive committees invariably consist of the most influential members of a department, you'll strengthen your relationships with people most likely to influence, in one way or another, your own future in the department. If chosen to serve on the executive committee, take it as a sign that you've already done something right, and seize the opportunity.

Courses and Curriculum: Most departments have some version of a "courses and curriculum" committee. The committee is in charge of evaluating periodically the soundness of the undergraduate and, in some cases, the graduate curricula—course requirements, distribution requirements, and so on. Especially during times when curricula are undergoing changes, the C&C committee can be a tremendously informative and practical unit on which to work. Not only will you gain a sense of the philosophies driving your department's approach to teaching, but you will also possibly have a direct say in what direction that approach will take in the near future and how your own ideas might be usefully incorporated into the program. You should ask more experienced professors in your department what they think about the C&C committee. In certain cases, you may be warned to stay away from such committees, since curricular work in some departments can be merely tedious and unproductive.

Search Committees: Involving yourself in your department's searches for new faculty members can be both a personally rewarding and a professionally beneficial experience. Especially if you are employed by a small college, one faculty hire can drastically affect the future course of the department. By participating in such searches, you can help to shape the department's future by hiring the sort of person you think would best address student and faculty needs—the sort of person you think would best help to improve the department. Share with your department head your willingness to participate in relevant searches on your campus. Be careful to choose your search committees wisely, though, since they can require a great deal of time and work.

College and University Committees: Often departments elect or appoint only their most experienced members for college-level and university-level committees. Nonetheless, opportunities do arise occasionally for assistant professors to serve on committees such as the faculty senate or the graduate faculty senate, almost always as departmental "representatives." My advice is to seize such opportunities. Such committees will give you a sense of how things operate at the highest levels, and you will meet people from departments other than your own (always good for your sanity).

SPECIAL CONSIDERATIONS FOR MINORITY STUDENTS/FACULTY

> Women and minority students have a greater burden because [institutions] want diversity on committees. . . . They do a disproportionate amount of the service work.
>
> *Chronicle of Higher Education*, 2003[4]

One of the most serious, and seriously under-discussed, problems in academe is the exploitation of minorities in service-related activities, a factor which may explain the higher attrition rates of minority graduate students. Because minorities constitute only 15 percent of the population of American citizens earning doctorates in the humanities,[5] it isn't difficult to figure out why so many problems occur; like so many industries in America, academe is trying hard to make up for a shameful past. Especially in an effort to recruit additional minority students and faculty members, many departments rely heavily on members of minority groups in their departments in order to advertise their diversity or at least their willingness to be diverse. Except in the cases of a few genuinely diverse universities—such as Rutgers and George Mason—such suggestions often amount to a form of dishonesty; one or two African Americans in a department of 50 professors does not a diverse department make. Now to be fair, many departments find themselves in something of a Catch-22 regarding such recruiting practices. To address and remedy the historical inequities that have led to the current crisis, higher education administrators must pretend that things are better on their campuses than they actually are; to welcome and make comfortable the best and brightest minority candidates, the same administrators must "use," and perhaps make uncomfortable, those members of underrepresented groups who have already joined their departments.

Were the difficulties of recruiting more minority candidates the only issue, the situation would probably not seem all that bleak. But minority members of academic departments are also routinely asked to act as departmental representatives on college and university committees, to attend luncheons and other functions during guest visits to their departments and campus job searches, and to serve as advisors to seemingly every minority student on campus, among other things. What is most disturbing, is that such pressures result directly and rather ironically in lower retention rates for minorities in higher education. I know personally three individuals whose academic careers were essentially destroyed, or at least seriously damaged, as a direct result of such race-based exploitation. Two of these persons were assistant professors so overwhelmed by the inordinate number of service obligations imposed on them that they were unable to build adequate research portfolios prior to tenure review. In one of the these cases, the professor asked her department head in her fourth year whether she might be temporarily relieved from two or three of these obligations so that she could be more productive in her scholarship. When she was told "no," she decided that it would be better to seek employment elsewhere than to be denied tenure. The third individual was a graduate student so burdened by service that he dropped out at the advanced ABD stage.

So what can you do if you should find yourself in such a predicament? There are no simple answers, but the following advice hopefully will be of some use to you:

1. Maintain a service diary similar to the one mentioned above. By keeping careful track of the hours you spend on service-related activities, you'll be able to demonstrate, should the need ever arise, just how difficult it has become for you to perform your research and teaching-related work.
2. Assess regularly your level of comfort about being asked to perform activities of questionable relevance to your career—whether they happen to be time-consuming or not. For example, African American art history students may feel pressure to participate in activities related to African American art even if they have chosen to specialize in Romanesque architecture. While such students may or may not decide to attend an upcoming lecture on an African American

painter, they would be justified in feeling exploited by a department that uses them to advertise the field of specialization. In such cases, the students would be well advised to articulate their feelings about the matter.

3. Maintain your integrity. Is your department asking you to present an impression of your university that you feel is misleading to minority candidates? If so, you may feel like your hands are tied. After all, your desire to be honest in no way suggests your insincerity as a recruiter. The best way to handle such a situation is simply to inform the relevant administrators that you will be an enthusiastic and friendly recruiter but that you are unwilling to distort reality for the sake of recruiting.
4. Make clear to your major advisor or department head any concerns you may have about the appropriateness of your service load. If necessary, be prepared to present your service diary as additional evidence of what's been happening. If you are a student, your major advisor should intervene for you in a legitimate case of departmental exploitation; if you are an assistant professor, your department head should respond reasonably and immediately to your complaints. If he doesn't, inform the dean of the problem and explain that you are resigning from X committee. Deans want to know how you are doing toward tenure, and if you bring them into the picture, you may well get support since they want you to qualify for it. If you voice your concerns only to be ignored even by the dean, then you should consider contacting either human resources or, if applicable, your local union.
5. Never forget your goal, which is to be tenured. Every well-meaning graduate student and professor feels bad about saying "no" to requests of one sort or another. Learning to say "no" from time to time, however, is one of the important lessons of academic life. Sometimes saying "no" to unreasonable service requests means saying "yes" to more important university obligations, such as good teaching and effective research. When you're feeling overextended and, at the same time, guilty about pulling back a bit from service, remember that you will be of no use to your department or university if you drop out of the program or lose your job. More important, you will have surrendered your own dreams and professional goals for the sake of an institution that should have treated you better.

Solutions for what we might reasonably refer to as a "crisis" in higher education—that is, the institutional exploitation of minority students and professors—must be addressed at the highest levels of university administration. Disciplinary organizations such as APA and ACLA should issue guidelines for departments; colleges and departments should offer counseling and mentoring programs where they are needed. Faculty should be educated about the problems with which their colleagues are contending on a regular basis. In the meantime, you can keep your head above water by staying focused on your own goals and being willing to communicate your concerns. Especially by establishing your willingness to participate in service-related activities with which you are comfortable, you will send a clear message that you are contributing positively to the overall life of your department, and you also will empower yourself to decide which service activities are appropriate and which ones are exploitative.

Conclusion

Before getting into my U-Haul, Ph.D. in hand, and driving away from Happy Valley, I asked one of my advisors whether she had any last-second advice for a soon-to-be assistant professor. She smiled and encouraged me to demonstrate "utter incompetence" on all departmental committees. She was joking, of course; utter incompetence would do little to help an assistant professor's tenure case. Her advice amounted nonetheless to a useful and pointed critique of the dilemma facing all competent persons: do well, and you will be asked to do more. I understood her point only because I had already accumulated considerable service experience as a graduate student. I also realized that the alternative service dilemma—do poorly, and you will be permitted to do nothing—was considerably more problematic. The benefits of establishing a positive ethos in your department extend way beyond placing yourself in a position to avoid job market or tenure problems; they have more to do with authorizing a credible voice—one that clearly will be heard as you begin to articulate your vision of what a university should be.

CHAPTER 12

THE JOB MARKET

John Guillory is correct, of course, when he claims that what graduate students want most is a job.[1] This book has aimed to give you the practical information and advice that you will need in order to earn that job someday. Ironically, one of the areas in which too many departments fail to provide proper training for their students is job market preparation. Since even the most qualified Ph.D.s will struggle to earn jobs if they aren't educated about how the academic market works, it seems appropriate that this book should conclude by outlining the process of applying for professorial positions in one of the most competitive job markets in America. In this chapter, then, I discuss the following issues related to job hunting in the humanities:

- Deciding when to go on the market
- Applying for appropriate positions
- Building a strong application
- Preparing for interviews
- Interviewing at conferences
- Interviewing on campus
- Accepting a job

In a single chapter, I can only begin to touch on the numerous, relevant issues that several other authors have discussed at book length. In addition to reading this chapter, therefore, you should be sure to attend whatever training sessions your department or university might offer, and you should read as many relevant books and articles as you can get your hands on. While several studies offer useful advice about the academic job market, the most thorough and practical of these is Kathryn Hume's excellent guide *Surviving Your Academic Job Hunt* (Palgrave, 2005), which you should make it a point to read prior to going on the market. In the meantime, here's what you'll need to know right away about the process of seeking a job.

WHEN TO GO ON THE MARKET

The vast majority of professorial jobs in the humanities are announced in the fall, filled in the spring, and begun the following school year. Generally speaking, the ideal times to go on the job market are after you've defended your dissertation or during the same fall semester in which you plan to defend it. As I discuss below, you might consider sending out applications earlier on in your career, but you will be far less attractive to potential employers as an ABD, which, by the way, is completely fair. Since most of the applicants will have defended by the time they apply for a job, university and college search committees usually make the logical decision to avoid the risks involved in hiring a person who *may* never finish her dissertation or whose delays will prevent her from working on bigger and better things. Such committees benefit greatly from an academic market in which supply far outpaces demand.

Let's consider for a moment the advantages *for Ph.D.s* of defending the dissertation prior to the first round of job interviews. First of all, you need to know now that job hunting is a full-time affair—an extremely time-consuming and mentally draining process. Many individuals who plan to defend in the spring semester, say, go on the market in the fall with excellent and seemingly realistic intentions, only to find out that their job market trials have prevented them from working on their dissertation for four months. Now they won't be able to defend until the end of the summer, at the earliest. If they don't manage to secure a job, they have nothing to show for stretching out their dissertation another four months or more. When such ABDs do succeed in earning a job, they face considerable pressure to finish their dissertations while contending with the stress of moving to a new location and trying to prepare—both mentally and physically—for beginning the new job. As David Chioni Moore has pointed out, "things can be [even more] problematic" for those new professors who fail to complete the dissertation in the summer before that first job begins: "your initial salary and rank may be lower, and, as history has often shown, it is difficult to complete a dissertation in the first year of a responsibility-filled full-time job."[2] Now, dealing with stress and inconvenience in exchange for a job might be worth the effort were it not for one additional factor: that ABDs who manage to secure jobs are in most cases doomed to question, perhaps for the rest of their careers, whether or not they allowed themselves to compete for the best possible jobs in their fields. That is, in the vast majority of cases, such persons are likely to be hired by less competitive institutions (again, why would a competitive one hire an ABD?), which raises the practical question: is your goal simply to obtain a job, any job, as soon as possible? Or is it to put yourself in a position where you might obtain a job that actually makes you happy?

Moore's excellent article on "Timing a First Entry onto the Academic Job Market" persuasively argues that since few job seekers succeed in obtaining a job their first time out on the market, it may be wise for students to make a "premature" run. His list of five reasons why a candidate should at least consider going out on the market early includes: (1) the opportunity to gain knowledge of how the job market works; (2) the opportunity to gain knowledge of what search committees want; (3) the opportunity to practice interviewing; (4) the opportunity to gain knowledge of one's own strengths and weaknesses; and (5) the possibility that one might actually obtain a job. Moore's basic point is that if you approach an early entry onto the market as little more than a practice run, you'll be more likely to succeed when it really counts—that is, when you are *really* qualified to compete for good jobs.

While I agree with Moore's basic argument and regularly encourage my own advisees to read his piece, I would advise you against making an early entry onto the market unless you are absolutely certain of three things: first of all, are you certain that you will be able to view an early entry onto the market as nothing more than a practice run? That is, are you really prepared to deal with the emotional damage and depression that can result from a failed job search? Second, are you able to say with a straight face that an early entry onto the market won't affect your ability to complete the dissertation on schedule? Third, will you be willing to turn down a job that falls short of your standards for what constitutes a *good* job?

If the answer is "no" to any one of these questions, then you should wait until you have defended your dissertation before going on the market. If the answer to all of them is "yes," then you might consider an early entry onto the market. If you do, make it a point to apply selectively. In fact, it might be wise of you only to apply for those jobs that seem most appealing to you. One of my good friends in graduate school decided to make an early entry onto the market (one year prior to his defense), but he applied only for three positions, and he considered all of them to be dream jobs. He was thrilled obviously when one of the search committees requested to interview him at MLA; even though he did not get the job, he reentered the market the following fall with a Ph.D. in hand, interviewing experience under his belt, and easily revisable application materials in his filing cabinet. While there are not always easy answers to the question, "when should I go on the market," your emotional personality, your organizational and time management skills, and your awareness of the potential dangers and advantages of an early entry should help you and your advisors to make an informed decision.

PREPARING EARLY FOR THE JOB MARKET

By reading this book, you've already begun to prepare yourself for the job market. By drafting early on in your career such documents as a teaching portfolio, CV, and dissertation abstract, you will prepare yourself yet further. Still, there are several practical steps that you can take in order to ensure that you are as prepared as possible:

Attend Your Institution's Job Training Sessions a Year Early: First of all, if your department offers no workshops for its job seekers, it's time to sit down for a conversation with the graduate director. Someone needs to complain. If your department is like most departments, however, it probably offers new and soon-to-be Ph.D.s at least some information about how to land a job. In a few institutions, job training is excellent. I'm happy to say that at UConn, for instance, we run annual meetings on "Writing the CV and Cover Letter," "Building the Teaching Portfolio," "MLA Interviews," "Campus Interviews," and in the spring, "The Politics of the Assistant Professorship." In addition, we arrange for at least one mock interview for each student on the market.

My advice is for you to attend *a year or more in advance* whatever workshops and informational sessions your department offers. You'll find it much easier to manage your time and your nerves if you enter the fall semester during which you'll be on the market with a nearly complete application. Remember that you will be finishing your dissertation and perhaps teaching a course or two as well, so there won't be much time for generating documents from scratch. Further, attending sessions a year in advance reduces the likelihood that you'll learn a month prior to the application deadlines that you were supposed to do something 2 years ago. In other words, early training helps you to ensure you've been and still are moving in the right direction.

Read Everything: For the third and final time, I would encourage you to begin reading *The Chronicle of Higher Education* and

InsideHigherEd.org 2 or 3 years before your entry onto the market. Only extraordinarily rare volumes fail to include *something* on the job market or the job seeking process. As important, numerous articles treat subjects important to assistant professors, which you'll find invaluable almost as soon as you are hired. Finally, these resources are the best source of academic gossip out there, which you can be sure your interviewers will expect you to know and will question you about at one point or another during an interview.

You should also reread this chapter and all of the more detailed studies out there on job seeking. As mentioned above, Hume's book (2005) is the most thorough and appropriate for humanities Ph.D.s, but you might also consult Showalter's *The MLA Guide to the Job Search* (1996), *The Chicago Guide to Your Academic Career* (2001), Sowers-Hoag and Harrison's *Finding an Academic Job* (1998), and Heiberger and Vick's *The Academic Job Search Handbook* (1996). While much of the material will seem obvious to you after a while, it will help you to approach common problems from multiple angles. Finally, if your department keeps interviewing diaries or notes from former students who succeeded in landing a job, seek access to them. When I was a graduate student about to enter the market, I benefited greatly from several first-hand accounts of what the process would be like; unlike all of the books I had read, these documents were written from the perspective of my classmates—who had received the same training and education I had. I found immense comfort in the idea that someone who had been in my place the year before was now working in a tenure-track job.

Attend On-Campus Job Talks: That is, when your department interviews candidates on campus for tenure-track positions, you should make it a point to attend the candidate's lecture and any other public events related to her visit. You will find it instructive to observe how different candidates perform during the course of an interview. You will learn as much, or more, from what doesn't work as from what works, and each separate lesson will help to shape how you will perform when it's your turn.

Participate in Searches: Occasionally lucky graduate students are asked to participate in searches for new professors. While such participation rarely includes attendance at convention interviews, it often does include the initial review of applications, organization of and participation in the campus interviews, and deliberation about the strengths and weaknesses of each candidate. Seize any opportunities you are given to serve on a search committee and, at the very least, volunteer to help out with campus interviewing.

Prepare for the Emotional, Physical, and Financial Costs of Job Hunting: I applied for almost 60 jobs the year I was on the market. Although I was willing to take a job in just about any part of the country, I was hopeful that I could stay on the east coast and so a Shakespeare and Milton job at the University of Connecticut immediately rose to the top of my list. I was thrilled four months later, therefore, when Connecticut offered me a job during my campus visit. Although I asked for a few days to consider the offer (see below, page 281), I knew in my heart that I would take the job. I got into my car an hour after my last on-campus meeting and prepared to make the six-hour trip back to State College. While gassing up the car five miles off campus, I called my wife and told her to chill that bottle of Dom Perignon we had been given at our wedding. Everything was grand. By the time I got home six hours later, though, I wasn't in a drinking mood. My temperature had risen to 104°, and I was beginning to hallucinate. Could this be real? The moment I had been working toward for six bloody years and now I was sick?

The reality, of course, was that I *was* sick because, for the first time in four months, I allowed myself to let go. I tell this somewhat disheartening story to all of my advisees because it manages to convey that despite how stressful and exhausting the job hunting process can be, it often has a happy ending. In a few days, I'd be able to walk again, after all, and even better, I had a job. My advice is that you do whatever it takes to prepare yourself early on for the emotional and physical stress that the job market causes.

You also should prepare for the financial stress it might cause. Especially if you wind up interviewing at a convention, to which you may have to fly, the total costs of your search—which includes travel, new suits, copying, postage, and so on—may be upwards of $2,000. Obviously, this may help you to decide whether it's worth it for you to make an early entry onto the market.

THE APPLICATION PROCESS

While unfilled academic jobs continue to be announced all year long, the main hiring cycle kicks off in the later summer and early fall. The "Careers" section of *The Chronicle of Higher Education* lists jobs in all humanities disciplines year round, though it is by no means an exhaustive resource. More useful are the various discipline-specific lists such as the MLA Job Information List, the CAA Job List, and the AHA Job List, which tend to be published in mid-to late September. While print editions of such lists are still available, most

The University of Connecticut, Storrs announces a search for a tenure-track, Assistant Professor in the literature of the English Renaissance (16th- and 17th-centuries) to begin fall 2005. Preference will be given to candidates with expertise in either prose or poetry. Teaching will include undergraduate Renaissance courses, including Shakespeare, and graduate courses in the candidate's area of specialization. Evidence of scholarly productivity and instructional excellence are expected for tenure. Candidates must have the Ph.D. in hand by August 20, 2005.

Salary: Competitive. The University of Connecticut actively solicits applications from minorities, women, and people with disabilities. Please submit a complete application, including a cover letter, CV, three letters of recommendation, and evidence of teaching experience (writing samples will be requested at a later date) to: Head, Department of English, RL Search, 215 Glenbrook Road, U-4025, University of Connecticut, Storrs, CT, 06269-4025 by November 10, 2004.

Figure 12.1 Sample job advertisement

candidates will access them online. The lists are categorized in a variety of ways—by theme, by period, by author or artist, and so on. Figure 12.1 shows what a typical job advertisement looks like.

This particular advertisement is instructive in terms of both its vagueness and its specificity. On the one hand, the committee makes clear that it is willing to consider applications from just about any Renaissance scholar; on the other, the committee reveals the department's need for a specialist in nondramatic Renaissance literature specifically. While Shakespeareans are encouraged to apply, they also are encouraged to make a case for their ability to conduct research on and teach topics other than drama.

The advertisement also lists two important deadlines, one being the date by which the Ph.D. must be "in hand" (this is a graduation date, not a defense date), and another being the date by which the application must be received. Know two things about application deadlines: most candidates will get their applications in just prior to the deadline; therefore, candidates who get their applications in early have a potentially significant advantage over the rest. Because search committee members are likely to spend more time looking over an application early on in the process, when it's in a pile of 10 applications, and far less time when they have as many as 200 other applications to review, you should do your best to beat the application deadline by several weeks.

Finally, job advertisements always inform applicants which materials they should submit to search committees. In what follows, I discuss the various components of a typical application.

The Standard Application

Regardless of whatever else you will be asked to submit, every single application you mail out will consist of three standard documents: a cover letter, a CV, and a dissertation abstract. While we have discussed the CV at some length elsewhere in this book (see pp. 69–75), the other two documents require brief attention here.

The Cover Letter

A cover letter is a two-page document highlighting what you can bring to your new department in the three areas of research, teaching, and service. Your cover letter should begin with a clear explanation of why you are writing. The first paragraph should convey your wish to be considered for a specific job (name the position since the department may be conducting multiple job searches) and explain how you learned about it. In this example, notice how brief is the writer's general introduction:

> Dear Dr. Spock:
> I am writing to apply for the assistant professorship in American ethnic literature advertised in the MLA *Job Information List.* My teaching experience and research in ethnic literature of the United States and Native American literature fit your job description well.

The second paragraph (and perhaps the third as well) of your letter should describe your dissertation project in the service of conveying not only the argument and content of a specialized research project, but also the range of expertise that you can bring to a department. A good dissertation description will make clear not only the merits of the dissertation as a dissertation, but also how the project is likely to be received beyond the dissertation stage, which is what employers really care about anyway. In this excerpt from a cover letter that I submitted to various search committees, I tried to build around my dissertation a sense of the project's marketability and its potential to produce yet more research (see also the appendix pp. 318–20 for examples of cover letters):

> An early version of Chapter One, "The Legacy of the Anti-Sports Polemic in Early Modern Prose," will appear in a forthcoming volume of *Prose Studies.* Part of the final chapter, "*Samson Agonistes* and the Politics of Restoration Sport," has been tentatively accepted by *SEL,*

> and the second chapter, on Shakespeare's *Henry VI*, has been revised and resubmitted to *Renaissance Quarterly*. As I prepare the manuscript for book publication, I plan to add a chapter exploring how Early Modern gender relations were influenced by the developing science we now call "exercise physiology."

The key is to show your potential employers your awareness that a good dissertation (and any other *past* achievements) will mean very little to them. A good dissertation that can be published as a book or six articles, however, might mean a lot.

You should devote at least one paragraph each to your teaching and your non-dissertation-related research credentials. You may wish to create two job letter templates, one aimed at "teaching schools," in which your teaching record would be highlighted more emphatically than your research, and one aimed at "research schools." In any case, the goal should be to demonstrate how the experiences you've accumulated as a graduate student prove that you will contribute immediately to the hiring department. Another paragraph should detail your service experience, and a final one should offer practical information should the search committee wish to interview you or simply collect more information from you. If you plan to attend a conference where interviews are typically conducted, then say so before closing. The cover letter is arguably the most important document you will write as a graduate student or recent Ph.D. Begin drafting the letter in the summer before your entry onto the market, and seek advice from your advisors, as well as the coordinator of the job training sessions at your university.

The Dissertation Abstract

By the time you are nearing completion of your dissertation, you will have written several abstracts of your dissertation, including the UMI abstract (see pp. 193–94). This is basically more of the same. Along with your job application, you should submit a one- to two-page abstract that highlights the topic, the argument, the range of content, and the contribution to existing scholarship of your dissertation. The appendix features two different types of dissertation abstracts (pp. 321–25). Whereas the first one appropriates a more standard form for abstracts—basically, a prose summary of the dissertation—the second reworks the form in order to address specific questions for which employers might want direct answers. What the second type of abstract gains by addressing the practical needs of employers it probably loses by violating traditional expectations of what such a document should look like. I wouldn't make too much of this; choose a form that best suits your needs and go with it.

The Dossier

Most employers request a dossier up-front, though some wait to review the standard application before doing so. The dossier consists of official transcripts and letters of recommendation, and it must be mailed either by your university's career services center or your department's graduate director or secretary. Because your goal should be to submit applications rather early in the fall semester, you'll want to set up the dossier no later than mid-September. What this means, of course, is that your professors must have at least one month prior to mid-September for writing their letters of recommendation. You should let them know sometime near the end of the spring semester that you will be on the market the following fall; sometime in August, you should provide them with a folder including an updated CV, a completed dissertation chapter or two, and a thank you letter specifying a clear deadline for their submission of the letter to the proper office. If your major advisor has not yet observed your teaching, schedule something for early in the semester (ideally, you'll do this in the previous semester). The importance of these letters can not be overstated, and it's your job to ensure that your professors will have what they'll need to write the strongest letters possible. Make sure your dossier includes at least three letters, arranged (by your advisor) in the order in which you would like them to be read, but feel free to include as many as five or six.

Additional Materials

If a school is convinced that you are worthy of an interview, they will almost certainly request additional materials from you. While some search committees demand such materials up front, along with the standard application and the dossier, most will request them by e-mail later on if they are interested in your application. Some of the materials you may need to submit include the following.

A Writing Sample

Most search committees will request an approximately 20-page sample of your scholarly writing. Ideally, you'll send potential employers a dissertation chapter that addresses the specific needs of the department as described in the job advertisement. That is, applicants for the UConn job advertised above would be better off sending a chapter on Spenser or Donne than a chapter on a Renaissance dramatist. When I was on the market, I worked out three different writing samples for three different types of jobs, one focused on prose writers, one on Shakespeare, and one on Milton. If you've published a relevant chapter or article, then you should send an offprint or a clean photocopy of it since search committees will respond much more enthusiastically to published work than to manuscript pages. If you send a manuscript that has been accepted for publication, make this clear in a header on the first page.

A Teaching Portfolio

Since we've already discussed in some detail how to create a teaching portfolio (see pp. 137–46), I'll simply say here that you might not always want to wait until a portfolio is requested before sending one off. Because I was concerned that my experiences in large research universities, as both an undergraduate and a graduate student, might affect how smaller schools looked at my application, I submitted portfolios to every teaching school to which I applied. I only submitted my portfolio to research schools upon learning that they wished to interview me.

Miscellaneous

Different colleges and universities will request any number of additional forms and documents ranging from affirmative action cards to teaching evaluations to syllabi tailored specifically to their curriculum. I was asked by one committee chair to submit a letter explaining my willingness to teach in an interdisciplinary Philosophy and Literature Department and proof of my ability to do so. For another potential employer, I was asked to confirm in writing that I had no objections to teaching in a Catholic institution. The point is that you should be prepared to generate any number of odd documents that you won't be able to anticipate.

You can never be sure what a given year on the market will bring. In order to see what you might be dealing with, have a look at the document shown in figure 12.2, which records one candidate's actual tracking of the process during a year 2000 search.

Your search may go very differently, or it may go similarly. The goal is to be prepared for anything.

Initial Round		
Number of schools applied to (out of about 80)	56	70%
Number of schools interviewing at MLA	50	89%
Number of research jobs (loosely defined)	27	48%
Number of teaching jobs (loosely defined)	29	52%
Number of schools requesting writing sample up-front	8	14%
Number of schools requesting dossiers up-front	27	48%
Number of schools requesting miscellaneous materials up-front	3	0.5%
Second Round		
Requests for writing samples (% of available schools)	15	31%
Requests for dossiers (% of available schools)	13	45%
Requests for miscellaneous materials	2	3.5%
Interviews		
Requests for MLA interviews	12	24%
Requests for phone interviews	2	33%
Campus interviews from MLA	4	33%
Non-MLA requests for campus interviews	2	100%
Offers		
University of X on January 25th		
X Tech University on January 30th		
Total costs of job search		
Approximately $1,600		

Figure 12.2 Job market application statistics

On the Market

The long process of applying for and securing a job usually begins sometime in August and ends sometime in March or April. In what follows, I break down the process into five separate but interrelated phases.

Pre-Application

Between August and about October 1, you'll need to establish your dossier and revise into flawless form your CV, cover letter, teaching portfolio, at least two writing samples, and a dissertation abstract. Start with about 15 copies of each of your writing samples and 25 copies each of the other documents. In addition you'll need to establish a filing system for organizing the hundreds of pages you're about to generate as a result of building your application and researching schools with available posts. Here's what to do:

Individual Job Files: Your filing cabinet should be set up to receive separate files for each college or university position for which you apply. Each file should include, at the very least, a materials checklist (see figure 12.3), a copy of the job advertisement, and any important materials you print out from the department's website.

School Name____________________________

1. Teaching school CV____ Research school CV____
2. Teaching school cover letter____ Research school cover letter____
3. Dissertation Abstract____
4. Abridged Teaching Portfolio____
5. Dossier______ Requested on______
6. Writing Sample #1____ Writing Sample #2___ Requested on____
7. Interview Scheduled for___________
8. Campus Interview scheduled for__________

NOTES:

Figure 12.3 Job market checklist

Materials Files: In your filing cabinet, preferably in a separate drawer, you should also set up individual files for each major application document: CVs, abstracts, and so on. You'll rid much of this stuff from your life once the initial round of application submissions is complete, but jobs continue to be announced throughout the fall, and often into the spring, so you'll want extras on hand at all times. Better to do a lot of copying up-front than to have to make copies every few days.

The Materials Checklist: Work out a form that will help you to keep track of your application materials and where they've gone. You will need about a 50 copies of the checklist, which should look something like the one in figure 12.3.

Notice how the job candidate accounts for all of the major documents and leaves room in a "Notes" section for atypical occurrences or requests. This type of form may not be the right one for you, but trust me, you'll need some kind of form for keeping track of what you send to whom, and when.

Dossier Services: Such services operate differently in every institution. Your goal at this stage should be to figure out how to make them work most smoothly for you. When I was on the market, I realized that it would not be efficient for me, every time I received a dossier request, to have to run across campus to career services in order to pay the $3.00 postage fee. Since the career services staff was more than happy to receive release requests by e-mail, I decided to put down $100 up front. This allowed me to send the staff each requesting college's mailing address by e-mail, which saved me considerable time, and forced me to visit the office only when my credit was running low. For the initial application submission, I made extra mailing labels for each of the 50 or so schools to which I was applying, and I asked the dossier services staff to place them inside my file. Now when they received one of my requests, they simply had to copy the file, stick a mailing address label on an envelope, and ship it off. You can't afford delays on the market, so you should do whatever it is in your power to do to keep things moving.

Post-Application

Most inexperienced job seekers believe that their work will end the moment they send off their applications. Nothing could be less true. Not only will you continue to search each week for job announcement updates, but you'll also need to send out new applications every time you find one. Further, schools will begin contacting you shortly after they receive your application in order to request additional materials. There are several other important steps you might take at this stage to prepare for interviews:

Know Thy Dissertation: Memorize 2-minute and 30-second oral summaries of your dissertation. The reason for doing so is because every interviewer you face will ask you the following question: "Can you tell me about your dissertation?" Since you'll want to keep most of your answers short, the 30-second version will be most important. Articulate succinctly and provocatively the argument, the scope, and the contribution to scholarship of your dissertation. End this summary by explaining that you'd be happy to provide more detail or answer any specific questions about the project, which may lead to an invitation to present the longer versions of your summary. At some point, you'll use those talking points that you've worked so hard to memorize.

Practice Interviewing: Obviously, you should arrange several mock interviews prior to a real one. If your department offers no interview training or mock interviews, ask your advisory committee members if they'd be willing to organize one. You might also consider organizing mock interviews with the other persons in your department who happen to be on the market.

Interview practice shouldn't only take place in formal settings. In fact, the majority of your practice sessions will occur when you are alone. During your runs, in the shower, or while you are watching television, you should be rehearsing aloud responses to the sorts of questions you might encounter in an interview. You will be able to construct a list of such questions by attending training sessions, reading job search articles and books (Hume's book, for instance, offers nearly 20 pages of commonly asked questions), and talking to your advisors and professors. Here's an extremely short list of the sorts of questions you're likely to face:

1. Questions assessing your level of interest in a particular type of college: When I was on the market, almost every single Liberal Arts school began the interview in the following way: "I see that you studied at Rutgers and Penn State, two big research schools. Why are you interested in a liberal arts school" or "Why are you interested in X College specifically?" Note that the subtly different second question requires a very different response. "How would you get any research done on a 4/4 teaching load?" "How well do you think you'd get on in a small town/big city?"
2. Questions assessing your pedagogical expertise and vision: "How do you approach survey courses in your field?" "How can you justify a whole course on Cervantes? Or on anyone for that matter?" "How would you justify using literature in a composition course?" "If you had to teach an entire course around one book written in the last 30 years, which one would you use?" "Name one pedagogical theorist who has influenced you and explain why." "Tell me about your two dream courses."
3. Questions assessing your practical skills as a teacher: "How do you deal with defiant or apathetic students?" "A student decides to research the Holocaust in one of your classes. She reads a number of works by Holocaust deniers and argues that the Holocaust never happened. The paper is written eloquently and quite flawlessly. How would you handle this situation?" "What are you views on mandatory attendance for college students?"
4. Questions assessing your understanding of curricular politics: "What is the role of composition in an English department?" "We have a required diversity course. I believe that we should keep it but John here thinks we should get rid of it. What do you think?" "Which courses should be required of philosophy majors and why?"
5. Questions assessing your research expertise: "How would you define the current critical paradigm in your field?" "Who are the

most influential art history critics of African-American painting right now?" "What do you think of Richard Helgerson's work on nationhood?"

6. Questions assessing your ability to handle direct challenges: "Isn't the use of film just another way to dumb down your students?" "I see from page 14 of your writing sample that you think X = Y. I find this problematic, to say the least. Can you defend your view?" "You claim to be a cultural materialist. I'm wondering why we really need another one of these?"
7. Questions assessing your knowledge of university politics: "How would you defend the institution of tenure to a skeptical person?" "I'm sure you've heard about the scandal brewing at X University? What is your take on the matter?"
8. Questions assessing your personal character and/or history: first of all, you should note that public schools are not permitted to ask you questions about your marital status, ethnic background, religious status, and so on. More realistic "personal questions" might include: "Why did you become an English major?" "Why do you wish to be a Spanish professor?" "What motivates you." "What are your strengths and weaknesses."
9. Questions assessing your future goals: "What do you plan to do with your dissertation now that you've defended it?" "Where do you see yourself ten years from now?" "What do you hope to bring to the university that hires you?"

This is a very limited sample of likely questions, but one that hopefully sends a very clear message that interviewing will require serious preparation. Know that interviewers also will expect you to ask them questions. Often this is a point where many interviewers screw up otherwise competent performances by asking for answers to questions they should already know (eg., "How many students attend your university?"). As Hume remarks, your goal is to glean as much useful information about the needs of the hiring college as your interviewers are willing to share; she recommends questions such as the following: "What areas would you hope that I might develop for you?" and "What are you hoping this hire can do for your department?"[3] Save questions about financial matters (salary, computer support, etc.) and spousal considerations until you have actually been offered the job (or someone else brings them up).

Begin Working Up Syllabi: Since a majority of questions will focus on how you teach or would teach specific courses, anticipate which courses you are likely to be asked about and work up a sample syllabus for each one. If you happen to be an expert on classical rhetoric, you should be prepared to answer questions about introductory courses, upper-level courses, and graduate courses on the subject. If an interviewer asks you how your graduate course on the subject would differ from your undergraduate course and you answer, "Uh, I don't know," the interview will effectively be over. But think how impressed your interviewers will be if you pull out several copies of a syllabus for a graduate course on "Classical Rhetoric" and respond, "*Here*'s how I would do it." You need not include every single part of an actual syllabus, but you should provide a basic course description, a list of required texts and assignments, and a course schedule (see appendix, p. 325). When I was on the market, I worked up about ten new syllabi, added them to the pile of seven or eight syllabi I had already used as a graduate student instructor, and I prepared myself to explain and present each one. About one third of them were graduate course syllabi. If this sounds like a lot of work, just remember that someone out there will be willing to do it, and for the same job you want so badly. If ever there's a time to over-prepare, it's when you're on the market.

Buy Your Interviewing Clothes: Professional suits are a must for interviews so you should prepare to spend some money. One suit with several shirt and tie options should do for men; women may prefer to bring several outfits since they are considerably cheaper than a good men's suit. A brief case or attaché bag is professional; a dirty old backpack is not. If ever there's a time to spend your money, it's when you're on the market. You can save money by buying the previous winter during the season's-end sales.

The Convention Interview Stage

Note that not all humanities Ph.D.s go through convention interviews in order to obtain jobs. Many are interviewed briefly on the phone and then brought directly onto campus. For the majority, however, the first round of interviews will take place at their discipline's annual convention: CAA, AHA, APA, MLA, ACLA, and so on. Most conference interviews occur late in December or early in January. Preparing yourself in the post-application stage for the generic questions you are likely to encounter during an interview will allow you more time after scheduling one to think specifically about the department that might decide to hire you. Here's what you should plan for:

Scheduling Interviews: Sometime in the late fall, usually between December 1 and 24, you may receive calls or e-mails from search committee chairs (or department secretaries) who want to interview you. Make sure that you have a working answering machine with a professional recording on it. Next to your phone you should keep a blank schedule for the period during which the conference will take place. For example, if the convention lasts from December 28 through December 30, sketch out a three-day schedule for the conference, beginning with an 8 AM time slot and ending with a 6 PM time slot. Figure 12.4 demonstrates what such a schedule might look like.

Now when schools call or e-mail to arrange for an interview, you can mark down the proper time on your schedule, which will also help you to prevent scheduling conflicts should you get more than one interview.

Some interview calls (some search committees choose to e-mail candidates) will be very brief and business-like, others conversational and friendly. You may find yourself talking with a search chair for five minutes or more. Regardless of whatever else you talk about, make sure you have answers to four questions before hanging up the phone or pressing send: "Which hotel or convention center room will the search committee be in? If the committee plans to interview you in a hotel," "whose name will the room be listed under?" "Which faculty members will be interviewing me?" "How long will the interview last?" Interviews usually are a half hour, forty-five minutes, or a full hour. Once all of your interviews have been scheduled, you can type up a master schedule, as shown in figure 12.5.

Preparing for the Specific Interviews: Begin to research more seriously each department with which you are interviewing. While you will want to know every person on the department's faculty and something about what each of them do, most of your research will consist of getting to know those members who will be present at the interview. If they have published books, read enough from them to gain a sense of their interests and influences. If they haven't, look for relevant articles. Read their biographies if available on the departmental web site, and check out their personal web sites. Becoming informed about each of your interviewers will make you feel like you know

Time	*Dec. 28*	*Dec. 29*	*Dec. 30*
8 AM			
9 AM			
10 AM			

Figure 12.4 Sample interview scheduling calendar

My Interview Schedule

Wednesday, December 27		
4:00	UColorado:	Marriot Wardman Park (2660 Woods Road / 502-329-2090). Jennifer Jones Suite
Thursday, December 28		
9:45	Southwestern:	Renaissance Suites (112 Connecticut Ave. / 206–347–3080). David Hasselhorf Suite
10:45	Westminster:	Grand Shalet (100 H St. NW / 206-583-1234). Joseph McCarthy Suite
1:30	La Salle:	Best Eastern Newington Suites (1111 Hampshire Ave. NW / 282–457–0565). James Earl Jones Suite
3:30	Idaho State:	Omni Hotel (250 Calvert St. / 205-234-0700). Natasha Kinski Suite
5:00	UConn:	Washington Hotel (191 Connecticut Ave. / 206-483-3000). Audrey Hepburn Suite
Friday, December 29		
9:00	Texas Tech:	Wyndham Trees Hotel (1400 M St. NW / 205-429-1700). Donald Duck Suite

Figure 12.5 Sample interview schedule

them, which will comfort you during the interview, show them that you've done your homework, and prevent yourself from putting your foot in your mouth. In order to help you memorize the most significant facts about a particular college and its search committee, work out cheat sheets for each interview (see figure 12.6).

Even so brief a document as this will help you to memorize key facts, and it will also help you to refresh your memory ten minutes prior to the interview.

At this point you might also consider tailoring relevant syllabi to the college's specific curriculum. For example, you might need to alter a 16-week course schedule to fit a 14-week curriculum. You definitely will want to change the title of courses so that they are consistent with the department's own language, and you may even want to include the appropriate catalogue numbers for each course. (See "customized syllabi packet" in appendix). I brought five or six custom-made syllabi to each of my convention interviews.

At the Convention: Conventions differ in size and style, but they all have several features in common. The largest of them is the MLA Annual Convention, which hosts thousands of scholars in the modern languages, comparative literature, and women's studies, among other fields and subfields. People often are shocked when they learn that most of the interviews conducted at such conventions take place in

U Connecticut Cheat Sheet

Job Description: Tenure-track assistant professor of Shakespeare/Renaissance literature. I know they want someone who can do Milton, even though not part of the advertisement.

Interviewers:
1) **Jean Marsden**: Director of Graduate Studies. Restoration and Augustan through Victorian women like Brontes. Books: *The Appropriation of Shakespeare* and *Shakespeare*, Adaptation *and Eighteenth-Century Literature.*
2) **Brenda Murphy:** American Literature, 1870 to present, modern drama, film and television. Has published around 7 books, mainly on *Death of Salesman*, but also on women playwrights and McCarthyism on television and film.
3) **Raymond Anselment:** Renaissance Literature to 1660 and Shakespeare. Has published 4 books, ranging from religious ridicule and English Civil War to this: *The Realms of Apollo: Literature and Healing in Seventeenth-Century England* (on medicine).

Where:
5:00 Washington Hotel (191 Connecticut Ave. / 205–483–3000). Jean Marsden Suite:

Department:
Faculty: 57 / Graduate Students: 96
Majors:?
Teaching Load: 2/2

Questions:
1) Murphy does film. Opportunities for Shakespeare and Film?
2) Opportunities for teaching Milton?

Figure 12.6 Sample interview cheat sheet

hotel rooms rather than in more professional settings. Some interviews take place at long tables or cubby-holes in ballroom warrens, in which multiple schools conduct interviews simultaneously. Considering how distracting such scenarios can be, the hotel room interviews are actually preferable.

Each room or suite will be arranged in a slightly different manner than the previous one. In some cases, three or four chairs will be set up in a semicircle. In certain suites, you will sit around a table. And in several instances, your interviewers actually will be sitting on a bed. While three is the typical number of interviewers you will face, it's not uncommon for candidates to be interviewed by a department's entire faculty, maybe as many as 15 people. I can remember vividly one interview in a small hotel room where 12 people took turns asking me questions. Sometimes graduate students will be present at the interview. Sometimes unidentified people will enter and exit from the room while you are being interviewed. The point is that you must be prepared for just about anything.

About ten minutes prior to your interview, call up the search committee chair on the hotel's house phone and ask for the proper room number, provided that it hasn't already been emailed to you. You may have very limited success using the various convention job center services, which supposedly offer information about room locations, and so on. The fact is that few schools register with these services and so you'll wind up standing on a long line for no reason. Though it would probably be smart to try the services center just in case, know that phones and e-mail are infinitely more reliable. Once you make that call, it's time to get on the elevator, remembering to breathe, and blow them away.

During the Interview: While actual interview protocol is far too complicated a subject for me to cover in a few pages, a few basic words are in order. I've organized the following, brief section into three basic categories: what to do in an interview, what not to do in an interview, and what to bring into an interview.

1. What to do: Your primary goal at a convention interview (other than not bombing) should be to demonstrate a sense of collegiality. In a scientific survey of "What Search Committees Want" at the interview stage, respondents ranked on a scale from one to six various criteria that clearly demonstrate the importance to committees of a candidate's interpersonal skills and basic personality traits. The findings broke down in the following way:[4]

Performance at interview with the search committee	5.51
Candidate's ability to relate well to students like ours	5.35
Candidate's ability to get along with other faculty	4.89
Candidate's personality	4.65

The fact that the search committee has requested an interview tells you that it already respects your work and believes that you are likely to make an excellent teacher and scholar. What the committee wants to assess, at this point, is what sort of a colleague you are likely to make. Is this the sort of person I can picture myself working with for 30 years? Is this the sort of person whom I would like to represent my university after I have retired? Is this individual likely to be difficult? Is she likely to feel comfortable on our campus and in our department? So while you spend most of your time preparing for highly specific questions about your research and teaching, remind yourself to practice answers that reveal your enthusiasm about the institution interviewing you and your passion about the academic enterprise in general.

One highly practical strategy would be to pursue the goal of making at least one of your interviewers into a partisan. Straight B + s, you should remember, will get you nowhere, but one A and two B + s might very well win you another interview. While you want to avoid turning off others by playing too directly to one interviewer, you should recognize your potential allies and seize opportunities to strengthen potential alliances.

2. What not to do: As the previous section would suggest, you'll want to avoid, above all else, seeming overly deferential or arrogant. While I can't help you much on the latter problem, the former can be avoided by reminding oneself that, in relation to this particular group of interviewers, one has never been and one never will be perceived as a graduate student. Once you enter that room, therefore, you must avoid thinking of yourself as a graduate student and your interviewers as professors. Think of yourself as a colleague, a fellow scholar and teacher in whom your interviewers have already expressed interest. By deliberately avoiding an unnecessarily deferential air, you'll not only seem more collegial to your interviewers, but you'll seem more relaxed, which will translate into a better interview. A good interview is like a conversation, not an interrogation or a formal Q&A session; by stamping out of your mind damaging presuppositions about your own status or rank, you'll free yourself to participate in, rather than merely respond to, one of the most important conversations in your professional life.

3. What to bring to the interview: Now it's time to use all of those materials you've been working up since the late summer. Here's how I would approach the difficult issue of what to leave in your hotel room and what to carry in your attaché bag to the actual interview:

a) Convention Folder: Your briefcase, which will feel like a bag of bricks at the convention, should contain a single folder with basic convention information: an interview schedule, name badge, hotel confirmation letter, maps, cheat sheets, and so on.

b) Research Materials: After learning those schools that would like to interview you, be sure to print important information off of their respective web sites: course schedules and descriptions, department handbooks, faculty profiles, mission statements, "Fast Facts," and descriptions of the campus and local surroundings. All of these things should go into a folder that already contains the job advertisement, a copy of your cover letter, and any other materials you received from the school (some departments will send you programs, brochures, etc.). In addition to the information about the school in the most recent Barron's and Peterson's Guides, these materials are likely to constitute your major reading material for at least a week before each interview (and certainly the night before). You'll also want to bring along a copy of your dissertation—which you might want to skim over at the convention—and whichever job market books you've purchased. There will be no need to bring any of these materials to the actual interview, of course.

c) Interviewer Folders: You might consider giving each individual interviewer a folder that includes two or three documents: an updated CV (should you land any publications or receive any other good news after the initial application submission), the abridged teaching portfolio, and a packet of syllabi customized to their school. About the latter, consider including 5–7 syllabi, ranging from elementary service courses to surveys to honors and graduate seminars (see especially pp. 270–71). Interviewers tend to be very impressed and grateful for these materials, especially when they have not previously seen your teaching portfolio.

4. Extras: Finally, carry in your briefcase a large file with non-customized, multiple copies of every syllabus you've made (see Appendix), an extra writing sample, and extra CVs. About the syllabi: Even if you give each interviewer a packet of custom-made syllabi, you may find yourself reaching into this larger file constantly in order to answer questions about *other* courses. Bring *at least* five copies of each syllabus to your hotel, and bring at least two copies with you when you go to each interview. It is important that you not impose copies of written materials—such as revised CVs or sample syllabi—on committee members during an interview without first inquiring about their interest in reviewing such documents. I recommend that you strategically choose appropriate moments in an interview to feel out whether your interviewers are interested in further evidence of your credentials: e.g. "Actually, I've worked up a syllabus on this subject, which I've also tailored to your departmental course offerings. Would you like to see it?" If committee members do not wish to burden themselves with more paper, you'll still benefit from being able to talk through the details of a course you've really thought through.

The On-Campus Interview

Usually it will be a week or longer before you hear whether one of your interviewers would like to bring you to campus. In the meantime, you should write a thank you letter or e-mail to each search committee chair (see pp. 325–26). Should you receive a call or an e-mail, you can expect to make your first campus visit as early as mid-January and, in rare cases, as late as April. The campus interview is in ways the most exhausting part of the job hunting process. Instead of facing three or four interviewers, on campus you'll face thirty, forty "interviewers," many of whom own titles like "Dean" and "Provost." Unlike convention interviews, which last an hour maximum, campus interviews can extend over several days. Rather than panicking, though, you should remind yourself that you've made it to the final stage of the process; now you're likely one of three candidates still under consideration for the job. In what follows, we'll discuss whom you are likely to meet on campus and what you likely will be asked to do.

Similar to fielding a convention interview phone call, you should be prepared to ask exactly what will be expected of you at your campus interview. Request a clear schedule of events and meeting times in the case that the search chair fails to offer you one first. Here's what you'll likely face on campus:

Interviews, and More Interviews: You'll meet with everybody but the dean's mother when you're on campus. Different departments and universities run different types of interviews, but you can be sure you'll meet with some, if not most, of the following people/groups: in smaller schools, the provost or the president of the college/university; the dean of the College; the department head; the associate department head; the graduate director; the entire faculty (or, at least, those members who show up); the executive committee; the tenure and promotion committee; the job search committee; a librarian; a group of graduate students; a group of undergraduate students; the staff, of which one member will process your travel papers and reimbursement forms. While all of this will be rather tiring, and while each interview requires a slightly different approach, you'll be so practiced an interviewer at this point that simple rhetorical adjustments and a lot of caffeine should get you through.

At some of these interviews, it's likely topics will come up that were off-limits at the convention. The department head, for example, might choose to tell you something about the salary, the benefits, or the teaching load. The promotion and tenure committee might talk to you specifically about what you'll need to do in order to earn tenure. And you might choose to ask questions about money available for travel to conferences, child care, and whether or not the department offers a junior faculty research leave. Keep in mind, though, that until you've been offered a job, it would be extremely unwise of you to admit any hindrances that might prevent you from accepting the job. For example, if you happen to be married to an academic for whom you hope to secure some sort of work at the university, hold off mentioning this fact until after an offer has been made. Only then would you have any leverage for negotiating terms.

Breakfasts, Lunches, Snacks, and Dinners: Something you'll be a little less used to are the various casual settings in which many of these meetings will occur. The trick to dealing with such settings is to remind yourself that there's nothing "off the record" at a campus interview. Everything you say will find its way back to the search committee chair or the department head. Don't allow the casual setting to cause you to relax, at least not to the point where you might let down your guard. Luncheons and such other events are great times to demonstrate your collegiality and to show off your personality, but they also should be regarded as just another part of the interview. Especially where alcohol is involved, be careful.

The Presentation to the Department: You will have to give some kind of presentation to the department, either of your research, your teaching, or both. Such presentations might last twenty minutes or they might last an hour. They might involve your reading of a dissertation chapter, or they might involve a discussion of your future research plans. They might be given before undergraduates and graduate students, or they might be for faculty members only. In one situation, I was asked to make a forty-minute presentation of my book (i.e., dissertation) research, with a five-minute explanation of how I might incorporate it into my teaching, and in another, I was asked to pretend that the faculty audience was really a class of undergraduates waiting to learn something about Voltaire. Ask the search chair exactly what she wants you to do, and follow her instructions carefully.

The Teaching Demonstration: Especially during interviews at smaller colleges, you may be asked to teach an actual class, which will be observed by several faculty members. For example, when the search committee chair calls you to request a campus interview, she may inform you that you'll be teaching one of Professor Plum's 9:30 sections of Western Civilization. Request a syllabus immediately. In some cases you'll be granted freedom to assign a certain text ahead of time and teach material with which you are comfortable; in others, you'll be told that you have to introduce the topic of the French Revolution or the text of *Sir Gawain and the Green Knight*. In either case, though, you'll want to extrapolate from the syllabus what the students are used to and try to deliver accordingly.

While the teaching demonstration tends to cause inexperienced job hunters more stress than just about any other part of the campus interview, it probably shouldn't. Not only do the observers understand the difficulty of the situation in which you've been placed, but the students do as well, and nine times out of ten they will rally around a likable, sincere teacher. The key to succeeding in this scenario will be your ability to demonstrate your enthusiasm for the material as well as your grasp of it. You won't necessarily want to stray much from your regular teaching practice, since doing so will feel unnatural to you, but you might try to sample over the course of the hour different techniques and skills that you regularly practice. For example, a class on the French Revolution would obviously require some lecturing. Especially at a small school, though, you'd want to show that you are able to move beyond lecturing and into a more Socratic or dialogic mode. Be sure to divide the class in a way that shows off your ability to handle both arts.

Finally, you should be careful to work out multiple "back-up" plans for unexpected alterations in your on-campus schedule. In some rare cases, candidates have been asked to prepare a 20-minute presentation only to be told later, on campus, that they have a full hour slot for their presentation. Explaining to your potential hirers that you only have enough material to fill up a 20-minute slot won't go over very well. My advice is that you over-prepare at least enough to be ready for such unpredictable occurrences. Should you actually wind up doing what you were asked to do in the first place, you may still be able to work in the extra material by explaining what you would have done had you been allotted more time.

Assessing Your Performance: As when you leave each hotel room at the convention, you'll spend quite a bit of time wondering how well you performed on campus. And just like at the convention, you'll be wrong most of the time. The simple fact is that you will rarely be able to tell how things have gone until you hear official word from a department head or search committee chair. One difference with campus interviews, though, is that in some cases people will let you know how you're doing on campus (when you're doing well, anyway). If they should be so kind, smile and thank them, and then force yourself back into interview mode. Should you be offered a job on campus, which is not all that uncommon, show surprise (it probably will be genuine) and act thrilled (it probably will be genuine), but do not accept the job on the spot. What you should say is something like, "I am thrilled that you wish to hire me, and I'm very enthusiastic about this job. I did promise my advisor/partner/children, however, that I would not accept a job without speaking to them first. Would you mind my taking a few days to have these conversations and to assess the details of the offer?"

Departments are bound to respect such requests, of course. Keep in mind that AAUP guidelines suggest that departments grant all job candidates at least two weeks to respond to an offer. You won't want to string anyone out unnecessarily, but what you gain by delaying is negotiating power should the offer be less than satisfactory. If you simply say "Yes!" and begin jumping up and down, there's little reason for a department head to agree to your request three days later to bump up the salary from $45,000 to $50,000. Never say yes until you have a formal offer in hand and you've discussed the matter with your advisors.

Negotiating and Accepting a Job Offer

Your advisors will help you to determine what's negotiable and what's not in a particular contract. They also will help you to determine how much leverage you might have as a negotiator. Salary is the most important issue for the majority of new professors, but you should also be careful to ask about several other matters: first, what kind of computer support will the department give you? Presumably, they will give you a computer and printer. Will the equipment be new? Will it meet your standards? Will the department be willing to purchase software for improving a less than ideal situation? Second, how much money will be available for travel to research libraries or conferences? Next, what type of work might the department be willing to find for your partner? Fourth, is the department willing to pay your moving expenses? Questions such as these are more than reasonable, and you'd be foolish not to ask them. What you don't want to do is be so difficult that you wind up turning off the department head—and every faculty member to whom she speaks about you—prior even to your arrival on campus. Nor do you want to bargain for perks—no service, significant teaching reductions, an RA—that will cause your new colleagues to resent you. Again, talk to your advisors about what's acceptable and what's inappropriate.

Once you feel entirely comfortable about the details of the offer, the only thing left to do is sign on the dotted line. At few moments in your life will you experience such feelings of relief or such a sense of personal achievement. The rest of your professional career will begin very soon, but as always, you should allow yourself to sit back and enjoy the satisfaction of knowing that you've made it through one of the most competitive and challenging processes you could have taken on. Be proud that you've placed yourself in a position to earn what all graduate students really want: a good academic job.

Afterword

A Future for the Humanities

I believe your generation of graduate students may prove the most important one since the World War II G. I. Bill. In a time of increasingly limited economic resources, higher educational institutions are transforming themselves once again. Whereas in 1944 massive investments in higher education were made to prevent a post-war economic depression, this time around it's been *cuts* that have followed a post-war recession. Ironically, the US government's most substantial recent investments have been back into the very institutions that caused our woes in the first place. Although the Obama administration's well-meaning federal stimulus plan promises billions of dollars to higher education, in the form of everything from Pell Grants to research funds, few states will be able to use this money for much more than filling holes opened by drastic budget cuts. At best, the money's purpose will be to stop our universities from bleeding. But from about 1945 until the turn of the century, America's colleges and universities were part of what we might reasonably view as an expansionist project, and its benefits for undergraduates, graduate students, and faculty alike were tremendous; disciplinary fields proliferated, academic opportunities increased exponentially—especially for women, working class citizens, and minorities—and intellectual opportunities allowed scholars not only to pursue their most specialized interests but also to redefine in myriad ways traditional methods of study dating back to the universities of late medieval Europe.

But the academic bubble has burst. Disciplinary fields are being merged or obliterated altogether, academic opportunities are decreasing with higher tuitions and fewer jobs for graduates, and intellectual ones are greatly threatened by increasing corporatization and reduced funding. And while I do not wish in any way to mitigate criticism for the bankers and business minds who've acted so unethically, I do think the burst in academe at least partly attributable to a certain laziness that set in on college campuses in the latter half of the twentieth century. Graduate programs which thrived in those years are especially vulnerable right now. But while it would be easy enough simply to paint numerous gloom-and-doom scenarios, I'd like to offer in what follows some preliminary thoughts about how graduate programs might *learn* from the current crisis and adjust practices in ways that will keep graduate students prepared for the challenges we're facing. My suggestions center around the pragmatist principle I've touted throughout this book, which is that greater transparency must characterize twenty-first-century graduate study: transparency about the structure and goals of our programs; about the epistemological assumptions underlying our field conventions; and especially about the professional applicability of scholarly research both within and beyond the academy. I argue that transparency in graduate advising and teaching can lead to the greater transparency—meaning accessibility—of graduate and professorial research in the humanities, which is crucial to our future wellbeing.

The market for academic research in the 1980s and 90s looks a bit to me like the American commodities market of the 2000s. In simple terms, the aggregate demand for more research from administrators and publishers motivated a sharp rise in supply. The increased number of available venues for publication, opportunities for grants, and openings for graduate school admission led universities to overinvest in the quantitative value of academic research and (cheap) human resources. Consequently, increased pressure on faculty and graduate students to produce more research led *in many cases* to a decline in the quality of that research. Even as concepts like "interdisciplinarity" thrived, more and more academic research became so hyper-specialized and idiosyncratic as to discourage communication, not only between academic disciplines, but also between academe and the wider public. This development was bad enough in itself, but when publishing houses began closing and grant opportunities began to dry up due to macroeconomic causes, disconnection and non-communication facilitated a bursting of the academic bubble. For any disciplinary unit whose mission is unclear to government and external funding agencies, the consequences have been costly. For humanities programs, whose importance has become unclear to so-called "hard" research programs such as engineering, the consequences have in many cases been dire.

Transparency at the curricular and advisory levels can lead to more widely accessible research for two reasons: the first has to do with the simple principle that knowing how a machine works allows one to take advantage of the full range of operations it is capable of performing; from first-hand experience, I can say that graduate students trained to understand the inner workings of the academic publishing industry, for example—from the economic to the philosophical factors driving its operations—are more likely to perceive in scholarly writing both weaknesses and strengths that will define, and help them to *re-fine*, their own approaches to professional writing. The second reason has more to do with the methodological awareness that program transparency can foster: students trained to think about *why* we do what we do, rather than just learn *what* we do, will better understand not only the value of answering practical questions about their own work but also the best strategies for doing so.

As I've been suggesting, at a time of economic crisis, at a moment when legislators and the general public regularly express suspicion about what the hell actually is happening on college campuses, researchers' ability to answer the "so what question" is crucial. Obviously, we should not assume that the persons to whom we have to answer will speak the same specialized language that our advisors or classmates have learned to speak. More people suddenly want to know what our work is about, some for the wrong reasons, but there might be a silver lining here. Whereas the formal demands of our research require us to focus very self-consciously on what our work is about and why it matters, the idea of having to explain both of these things to an angry parent, a biology graduate student, or a dean of the college (who is responsible for allocating resources) allows us to take nothing for granted; it forces us to be extraordinarily adept at understanding the limitations, the fundamental assumptions, and the highly particular nature of the knowledge set we possess. Our fantasy of operating within our own little pastoral academic worlds can be quite appealing, of course, because it allows us to be relatively lazy. Generally speaking, I feel no need to explain to another English professor why *King Lear* is an intricately constructed, philosophically valuable, or even potentially life-changing work of art. But the moment I'm asked about a Shakespeare play by a skeptical pharmacy major or a CPA seated next to me on a plane, I am forced to begin doing some really difficult work. Since I've embraced my own doubts about what I do, however, instead of simply ignoring them, I should be able to provide answers.

The challenge for your generation, then, beyond a more willing interrogation of your own research practices, is for you to begin thinking about both interdisciplinary research and cross-disciplinary communication in more pragmatic ways than most of us have in the past. A quick glance at almost any recent issue of the *Chronicle* reveals the costs, especially for humanities and arts scholars, of failing to do so, since university budget cuts tend to target such programs disproportionately.

The current crisis also reveals the costs—and the sheer irony—of a governmental and institutional dispriviledging of the humanities. Even as the decline of economic and ethical fundamentals follows from the pursuit of higher and higher returns; even as the opportunities of an entire generation are threatened by government and corporate corruption; the humanities' potentially salutary influence is being lessened. I recently watched in bafflement as my own department was forced to cut a percentage of its budget almost twice as high as the total percentage of cuts in the College of Liberal Arts and Sciences. I was similarly baffled by the silence with which we acquiesced. Do we no longer have a voice? Or are we simply lacking the necessary confidence to make people listen? There's further irony, of course, in the fact that programs that teach oral and written literacy are being attacked when the decline of communication skills has been such a large contributor to the current crisis. For example, law schools—always a priority of academic institutions—are calling for more philosophy and language majors at the same time that resources allocated to training such majors are being cut. In order to counter such destructive economic measures, scholars across humanities disciplines need to come to terms with and learn to articulate the ways in which their fields directly inform and rely upon the health of other disciplines—and for no reason more important than that universities are massive organic units that function most efficiently when all their constituent parts are healthy. It's easy in times such as these for scholars to think only about protecting their own disciplinary turf, in part because we've allowed ourselves for so long to be embarrassed by important questions about what we do in and for the world.

Since I've been suggesting in large part that graduate study in the twenty-first century must be more pragmatic, I want to stress yet again that being pragmatic means in no way compromising the quality of the research we do, in part because, as I argue above (see pp. 11–17), the practical/intellectual dualism is as false a construction as the scholar/professional or the academe/real world dualisms that have led to this erroneous idea in the first place. Researchers can no longer afford to accept popular culture definitions of the merely "theoretical" or the "academic" (as in "that's merely academic"). In times of shrinking resources and growing skepticism, the *merely* or the unnecessary things will have little chance of survival.

In concluding it is worth remembering that in the long history of western higher learning, recommendations that pragmatism and intellectualism be wedded are hardly innovative. Plato's famous parable of the cave in *The Republic* focuses, in fact, on the obligation of a free society's most enlightened citizens to communicate what they have learned even to the state's least enlightened individuals. In that story, Socrates encourages his student to imagine a cave with only one opening. Inside, with their backs to the opening, are men who have been imprisoned since childhood, chained so that they can only stare at the rear cave wall. Outside, at some distance, a fire burns, which casts shadows on the wall. Socrates describes how the prisoners, seeing the shadows before them and hearing the voices associated with them, will come to assume that the shadows themselves are the *real* things. Socrates then asks us to imagine that one of the prisoners escapes from the cave. At first, he is blinded by the light of the sun and refuses to look at it, a struggle suggestive of the difficulty of renouncing one's former beliefs and embracing greater truths. But the enlightened individual eventually comes to terms with the real world and rejects the shadows, and the question turns to what the former prisoner and now free man should do next. Should he request a sabbatical? Require a teaching reduction? Socrates argues that the republic's best minds should be prevented from "remaining in the upper world, and refusing to return again to the prisoners in the cave below and share their labours and rewards."[1]

What I find most appealing is that Plato doesn't allow us to naively imagine the return to the cave. Socrates asserts that in the world of knowledge, the good appears last and is perceived only with great difficulty. The man who tries to convince the prisoners that what they see before them is merely an illusion will be called a liar, and he will be resisted or even killed. But his duty, nonetheless, is to return, explain what he has learned, and try to convince others that it is crucial both to their health and that of the state.

How transparent shall I be here? Are we more like the escapee who remains in the upper world refusing to return to the cave? Or the escapee who returns but lacks the ability to convince others what they most need to know? Or are we like the prisoners themselves, mistaking the shadows on the walls of the ivory tower as more real than the sun shining outside? Can we be more like the character Socrates, aware of the importance of what we know, and of the difficulties of communicating it to others, but still willing to try?

Such willingness to return may be less an option for generations of researchers following the so-called "Great Recession." Research in a vacuum will not be easy to accomplish in coming years. Your burden will be to convince others—whether in academe or in the workplace beyond the academy—that what you have discovered can impact their understanding of the world and ability to function in it. But beyond these practical necessities, the return to the cave is long overdue, and we might view it as a check on the sort of excesses that have allowed and sometimes encouraged us to cease communicating its importance to non-specialists. Partly for this reason, I have also suggested that it is the job of your graduate programs and advisors to foster meta-professional conversations geared toward the demystification of academic processes and disciplinary practices. Such conversations should reflect an interest in the realities students will encounter outside of the cave; they should be more respectful, even encouraging, of practical questions and sensible doubts and less focused on elite socializing practices. Conversations about everything from the value of work performed by a discipline to the "how to" side of research methods and professional development will not only help graduate students do more informed research but also develop more sophisticated ways of defining and disseminating the results. Such conversations enable clearer cross-disciplinary communication and more productive exchanges between the academy and the world beyond. Promoting greater self-reflection and definition, these new research models openly confront, rather than running away from, self-doubt, which has for too long been permitted to cast his shadow on the utter practicality of our intellectual heritage.

In thinking more about how to make ourselves heard, in other words, we might take a lesson from the discussion of written voice above (see pp. 108–14). And I think this means focusing especially on the ways in which our communal voice, like our individual ones, has been hampered by the burden of the past. For years, we in the humanities have spoken of the "intrinsic value of the liberal arts" and of other similar Renaissance ideals which, whether we believe in them or not, no longer carry weight within the current higher educational paradigms—which, lo and behold, are focused largely on the pursuit of greater returns, or science models that apply only awkwardly to what we do. If we want to be heard, we must focus less on the *legacy* of the humanities and more on how the sort of work we do as humanities scholars and artists can actually alter conversations about the *future*. And *some* of the extraordinarily practical things we do include: thinking critically and humanely; redefining our present and imagining our future by understanding and articulating our past; refining both our written and conversational voices so that our ideas apply outside of the academy; and of course, teaching our students how to do all these things. We must seek, in short, to reaffirm our place in a larger conversation that so badly needs our collective voice.

Appendix: Professional Documents

Sample CV

Sean Christopher

965–3 Southgate Drive, Little Town, MA 10891
• Phone (819) 235–1888 • E-mail: sc7@fu.edu

EDUCATION

Doctor of Philosophy in English, Fake University, December 2004
Master of Arts in English, Fake University, May 2000
Bachelor of Arts with Honors in English, *summa cum laude*, Fake Undergrad University, May 1998

DISSERTATION

Narrating Prisoners in the Victorian Novel
Scholars have long depended upon Michel Foucault's discussion of the surveillant Panopticon to shape their discussions of the novel and the prison. I argue, however, that the real Victorian prison was based upon a model of separate confinement, self-reflection, and self-narrated guilt that has other implications. The Victorian novel's relationship to the prison is most clear not in discussions of surveillance and social power but rather in investigations of the guilty autobiographical discourse that the prison both shaped and required. My project analyzes the ways in which Charles Reade, Wilkie Collins, Marcus Clarke, Charlotte Brontë, and—of course—Charles Dickens each wrote

For legal reasons, these documents—like many of the in-chapter figures— have been slightly fictionalized. Names of authors, supervisors, and schools have all been altered. Article and dissertation titles have also been changed, though most publication information remains unaltered. I have included some of my own professional materials in these appendices, not because of my massive ego but rather because I do not need permission of any sort to do so. Though most of these materials were generated by students of language and literature, they are applicable across the humanities.

novels reflecting and recreating this separate prison. These authors invented a body of literature that raises crucial questions about narrative authority, psychological exposition, and the private self. (Please see the enclosed abstract for a full discussion.)

PUBLICATIONS

Articles

"W. H. Auden and the Meaning of Shakespeare." Forthcoming in *South Atlantic Review* 66 (2005): 1–19.
"Narrating American Prisoners." *Journal of English and Germanic Philology* 99.1 (2003): 50–70.
"Dickens, The Past, and the British Prison." *Dickens Studies Annual* 29 (2001): 17–39.
"Nature's Perilous Variety in Nineteenth-Century Poetry." *Nineteenth-Century Literature* 51 (1999): 356–376. Reprinted in *Nineteenth-Century Literature Criticism.* Detroit: Gale Research, 1998.

Reviews

George B. Palermo and Maxine Aldridge White, *Letters from Prison: A Cry for Justice. Crime, Law and Social Change* 30.2 (1999): 297–298.
Richard Mowery Andrews, ed., *Perspectives on Punishment, an Interdisciplinary Exploration. Crime, Law and Social Change* 29.1 (1998): 83–84.
Seán McConville, *English Local Prisons, 1860–1900: Next Only to Death. Crime, Law and Social Change* 28.2 (1997): 180–182.

CONFERENCES

"Making Progress: Technologies of Narration in *Bleak House.*" *South Central Modern Language Association.* Tulsa, OK, scheduled for November 2001.
Panel Chair, "Victorian Contamination and Colonial Dis-ease." *Central New York Conference on Language and Literature.* The State University of New York, Cortland, NY, scheduled for October 2001.
"Narrating the Cell: Dickens on the American Prisons." *Southern Conference on British Studies.* Georgia State University, Atlanta, GA, 1997.
"From Spain to 'Oxford'." *Twenty-First Annual Colloquium on Literature and Film.* West Virginia University, Morgantown, WV, 1996.
"Nature's Perilous Variety in 'Goblin Market'." *Central New York Conference on Language and Literature.* The State University of New York, Cortland, NY, 1995.

TEACHING AND RESEARCH POSITIONS

Lecturer (full-time, 3–4 load), The Fake University, 2003–04
Teaching Assistant, The Fake University, 1998–2003
Research Assistant to Christopher Clause, The Fake University, 2000–01
Editorial Assistant, Faculty Senate Office, The Fake University, 2000–01
Edwin Erle Sparks Fellow, The Fake University, 1998–99

COURSES TAUGHT

British Literature from 1798 to the Present (1 section, enrollment 35)
English curriculum core course covering major authors from the Romantic period to the present.
Traditions in English Literature (1 section, enrollment 40)
Survey of English literature from medieval to the present for majors and nonmajors.
Reading Nonfiction (1 section, enrollment 35)
Investigation of the major forms and functions of English and Anglophone nonfiction prose.
Reading Drama (1 section, enrollment 35)
Study of dramatic genres and conventions from Sophocles to Stoppard.
Understanding Literature (1 section, enrollment 60)
Introduction to literary genres, themes, and devices for predominantly nonmajors.
American Comedy (1 section, enrollment 90)
Survey of American comedy from 1700 to the present for major and nonmajor students.
Honors Freshman Composition (2 sections, enrollment 24)
Honors version of Fake University's rhetoric course for first-year students.
Rhetoric and Composition (6 sections, enrollment 24)
Required writing-intensive course for first-year students.
Business Writing (10 sections, enrollment 24)
Required writing-intensive course for upper-level business majors.
Undergraduate Advising
Serve as one of four academic advisors to Fake University's 500 English majors.

HONORS AND AWARDS

Henry Bruner Young Excellence in Business Writing Award, 2003
English Department/E.G.O. Outstanding Teaching Award, The Fake University, 2001
Edwin Erle Sparks Fellowship (full-year award), The Fake University, 1998–99
College of Liberal Arts Travel Grant (special award), The Fake University, 1999

Departmental Travel Grant (special awards), The Fake University, 1998, 1999, 2001
Humanities Fellowship (annual awards), The Fake University, 1999–2001
Phi Beta Kappa, Fake Undergrad University, 1997

AFFILIATIONS AND SERVICE

South Central Modern Language Association, Member, 2001–Present
Modern Language Association, Member, 1998–Present
South Atlantic Modern Language Association, Member, 2000–01
Composition Program Teaching Awards Committee, Spring 2000–01
Modernist Studies Association, Assistant to the Conference Directors, 1999
Composition Program Teaching Series, Guest Speaker, 1998
English Graduate Organization, Masters Student Representative, 1995–96

REFERENCES

1. Chris Clause, Professor of English, The Fake University
   e-mail: cc@fu.edu office phone: (819) 863–9582
2. Philip Henkers, Distinguished Professor of History, The Fake University
   e-mail: ph@fu.edu office phone: (819) 863–8946
3. Michelle Aneker, Associate Professor of English, The Fake University
   e-mail: ma@fu.edu office phone: (819) 863–9583
4. Robert Loggins, Associate Professor of English, The Fake University
   e-mail: rl@fu.edu office phone: (819) 863–0283

All references may also be contacted by regular mail at: [Department]
The Fake University
College Park, MA 16890

SAMPLE COMPLETE SYLLABUS

English 666.01: Topics in Literature
Literature Goes to Hell

T/Th: 9:30–10:45 / 247 Pandemonium Hall

Professor: Dr. John Milton
Office: 232 Pandemonium
Phone: 486–4666

Office Hours: Tuesday, Wednesday, Thursday: 11–12:30
E-mail: John.Milton@Great.university.com

Course Description: This course explores literary renderings of hell and underworld figures, especially Satan, and analyzes the social, psychological, and philosophical functions of hell and devils within particular literary works and actual historical contexts. We will spend a good deal of time pondering the gradual transformation of the concept of hell, from an external, absolute physical space into a relative, internal condition. What psychological, social, and artistic functions has the concept of hell performed throughout history and especially in our literature? What might the gradual internalization of hell (as well as the occasional reemergence of absolutist definitions of evil) teach us about our own position in this larger intellectual and social history?

Texts: (please use editions ordered for class)

Bible (handouts)
Ovid, *Metamorphoses* (handouts)
Dante, *Inferno*
Marlowe, *Dr. Faustus*
Milton, *Paradise Lost*
Pullman, *His Dark Materials* (trilogy)
Rilke, "Orpheus, Hermes, Eurydice" (handout)
Polanski, *Rosemary's Baby* (film)
Scorsese, *The Last Temptation of Christ* (film)
Lewis, *The Monk*
Twain, *The Mysterious Stranger*
James, *Turn of the Screw*
Rice, *Memnoch the Devil*
Aeschylus, *Prometheus Bound*
The Book of Job (Mitchell trans.)
Virgil, *The Aeneid* (handout)
Homer, *The Odyssey* (handout)

Requirements:

Class grade (preparation, participation, attendance)*	15%
8–10 reading quizzes	25%
Take-home midterm paper	20%
Film review	20%
Final exam	20%

Office Hours: Office Hours will be held in 232 CLAS during the following times: Tuesday, Wednesday, and Thursday between 11:00 and 12:30. I urge each of you to take advantage of the opportunity to introduce yourself to me, to ask any questions you may have, to discuss future or current assignments, or to seek private instruction on specific problems with which you might be wrestling. I like students, I love teaching, and I promise that I don't bite. There is a too often unrecognized but undeniable correlation between students who tend to use office hours and students who tend to be successful in college. I also encourage *professional* communication through e-mail.

* The professor does not take kindly to excessive absences. Because your understanding and engagement of the material depends on your presence in our class, I will take absences seriously. More than two, for any reason, is inexcusable. Furthermore, excessive absences will affect your quiz and exam grades since the latter will test skills we will develop in class. Preparation includes completing homework and in-class assignments, and having read the texts to be discussed in class. Please make it a point to turn off your cell phone before class begins.

Plagiarism: It goes without saying that you are responsible for citing any words or ideas that you borrow. Using material from the so-called Internet Paper Warehouses constitutes a form of plagiarism as serious as using someone else's paper (and is easy to discover). Plagiarism demonstrates contempt for your instructor, peers, and the purposes of liberal education. **If you are caught plagiarizing, you will automatically fail the course for violation of the student code and be referred to the dean of students for judicial affairs**. If you are uncertain as to what constitutes plagiarism, please consult the English Department's policies guide or see me outside of class.

Quizzes: Quizzes are designed to test your basic reading comprehension skills; they are not designed to trick you. My sense is that you will do absolutely fine as long as you read each day's assignment carefully. A typical question might look like this: *Why doesn't Romeo receive the Friar's letter in Mantua?* Easy, right? Please note that missed quizzes cannot be made up unless you can show me an official University excuse for your absence (i.e., official athletic event, serious illness, etc.). Everyone may automatically drop one quiz grade. If you take *all* of the quizzes, you may drop one additional quiz grade. To dissuade lateness, I will give *most* quizzes in the first five minutes of class; I will not repeat question #1 of a quiz after I've moved on to question #2. Please be here on time.

Papers (including midterm and film review)
Goal: Your primary goal is to offer a clear, concise argument (claim and basis) about the text that you are analyzing, and to back up that argument with evidence, quotations, examples, and so on. Noting the often subtle distinction between analysis and summary is key to your success. The goal is not to explain what a book is about or how it is put together (summary)—rather, the goal should be to contribute something new or original to our understanding of the text. Remember that more is less in such a short assignment. Go deeply into one issue rather than shallowly into multiple issues. Dissect.

Example: A summary paper might be set up in the following way: *In* King Lear, *Shakespeare tells the powerful tale of a man who splits up his land among his three daughters. He asks each one to tell him how much she loves him. The first one answers . . . etc.*

An analysis, on the other hand, might be set up like this: King Lear *suggests that human love and compassion is the only life affirming force structuring our universe* (claim). The remainder of your paper would focus on supporting this claim through textual examples, close-reading, and quotations.

Whereas the first paper merely traces or summarizes the content of the play, the second offers a commentary on the play's argument, message, or significance. One is descriptive. The other is argumentative.

You should expect to begin this class at a certain level of expertise and to leave it at a more advanced one; you should not be surprised, therefore, to receive extensive, *critical* feedback on your first few written assignments. The idea is to provide you with the information you will need to make changes and to become a better writer. **See also the handout on Writing Guidelines and take it seriously**.

Film Review: Although your film review is officially due on the last day of classes, you may choose to get it done and turn it in much earlier. You may write on any of the three films we will be viewing as a class (*The Omen, Rosemary's Baby*, or *The Last Temptation of Christ*) or any of the following films: *Dogma, The Exorcist, Pleasantville, Bram Stoker's Dracula* (directed by Francis Ford Coppola), *The Devil's Advocate, Hellraiser, Angel Heart*. Please note that you will be quizzed and tested only on the three films we will highlight in class. More information on the review is forthcoming.

Midterm Paper: Information forthcoming in late September.
Final Exam: Information forthcoming toward end of semester.

Schedule

T	Aug. 31	Introduction to course; The Dionysian Principle and History of the Devil: Background Lecture #1
Th	Sept. 2	Introduction continued; The Old Testament: Genesis: 1–3; Book of Job
T	Sept. 7	The Roots of the Satanic: Gospel of Truth (HO); Aeschylus: *Prometheus Bound*
Th	Sept. 9	The Roots of Hell: excerpts from Ovid's Metamorphoses: The Rape of Persephone/Proserpine, Orpheus and Eurydice; excerpts from Virgil's The Aeneid; excerpts from Homer's The Odyssey; Rilke: "Orpheus, Eurydice, Hermes"
T	Sept. 14	The New Testament and Apocrypha: Background Lecture #2; Dante, *The Inferno*: 27–102
Th	Sept. 16	Dante, *The Inferno*: 103–196
T	Sept. 21	Dante, *The Inferno*: 197–288
Th	Sept. 23	Marlowe, *Dr. Faustus*: Acts 1–3
T	Sept 28	Marlowe, *Dr. Faustus*: Acts 4–5
Th	Sept. 30	Milton, *Paradise Lost*: Books 1–4
T	Oct. 5	Milton, *Paradise Lost*: Books 5–8
Th	Oct. 7	Milton, *Paradise Lost*: Books 9–12
T	Oct. 12	**Midterm Paper Due**; Lewis, *The Monk*: 1–104
Th	Oct. 14	Lewis, *The Monk*: 104–203
T	Oct. 19	Lewis, *The Monk*: 203–293
Th	Oct. 21	Lewis, *The Monk*: 293–363
T	Oct. 26	Twain, *The Mysterious Stranger*
Th	Oct. 28	James, *Turn of the Screw*: pp. 1–48
T	Nov. 2	James, *Turn of the Screw*: 49–85
Th	Nov. 4	Rice, *Memnoch the Devil*: pp. 1–160

T	Nov. 9	Rice, *Memnoch the Devil*: pp. 161–314
Th	Nov. 11	Rice, *Memnoch the Devil*: 315–434
T	Nov. 16	**Film Review due**; Read Pullman, *The Golden Compass*, 1–178
Th	Nov. 18	Pullman, *The Golden Compass*, 179–351
T	Nov. 23–25	**Thanksgiving Recess: No Classes**
T	Nov. 30	Pullman, *The Subtle Knife*, 1–146
Th	Dec. 2	Pullman, *The Subtle Knife*, 147–288
T	Dec. 7	Pullman, *The Amber Spyglass*, 1–225
Th	Dec. 9	Pullman, *The Amber Spyglass*, 226–465

Final Exam Time and Location TBA

CONTENTS PAGE FOR SAMPLE TEACHING PORTFOLIO

Abridged Teaching Portfolio
Professor Georgio Spumante

Table of Contents:

TEACHING PHILOSOPHY

My basic goal as an English instructor—based upon a philosophy I call meta-pedagogy—has been to make students aware of the educational process itself. Students are encouraged to become active participants in the construction of course syllabi, organization of class activities, and the conveyance of knowledge. They are encouraged to consider the implications of educational policy making and pedagogical presentation so that they might become more critical of the practices that affect their own acquisition and use of knowledge. I have focused on helping them to strengthen their convictions and stressed the importance of articulating those convictions in a variety of settings.

My experience has taught me that students often perceive educators not as people working to help them, but as obstacles or stepping-stones between them and their futures. I've come to realize that such (erroneous) perceptions are partially the result of their detachment from or nonparticipation in the educational system. Most students go to class, take their tests, complete their core requirements, and fill out their evaluations because they are asked to do so but not because they understand the reasons for doing so. However interested they may be in knowing those reasons, they are often conditioned not to ask about them, not to question the purpose or efficacy of traditional or nontraditional pedagogical methods. I have been impressed by the positive reactions of students once they are comfortable enough to ask these "forbidden" questions. For example, the first question I tend to be asked by writing students is "Why do we have to take these classes?" Several years ago, my response was typical of the unsatisfactory answers that are usually given: "Because every job requires written communication skills, etc." Now I assign interview papers that each student must complete. The student must arrange for an interview with a person in her prospective field (a dean, employee, professor, etc.). She must explain to the interviewee the class she is taking, and then she must question how it will be useful down the road. Without exception, students return to class after the interview more determined to work and appreciative of the concrete answers they've discovered.

I have embraced an interdisciplinary, multi-media approach to teaching in order to stress the connections between fields of knowledge that students often perceive to be unrelated. For example, in "Introduction to Shakespeare," we move from an in-depth examination of each play to musical and artistic reconstructions of Shakespearean drama such as Mendelssohn's Overture to *A Midsummer Night's Dream* and Henry Fuseli's painting of the same title. My classes integrate music, film and television clips, and trips to local art collections and playhouses to stress the complex pervasiveness of ideology and the exciting inter-connections between cultural media. Students begin to see knowledge as dynamic and alive, not fixed and static.

In conclusion, I admit that my greatest fear about meta-teaching is that I will be unable to maintain enough authority to conduct an effective course. After all, my courses teach students how to be critical even of me. I've learned that the fear is unnecessary. By focusing students on the learning process, I help them to understand the highly complex factors that influence my assessment of their performances. They begin to feel as though they can understand and control these factors as well. Grades are less frightening as a result. They become markers on a quite accessible pathway to improvement and success. As a teacher, I have tried to empower students while maintaining rigorous standards of excellence.

Summary Report on Undergraduate Student Evaluations

English 267.01 "Literature Goes to Hell" (Fall 2003)
40 Students
Overall Mean: 9.5 out of 10
Overall Median: 10.0

Student Comments

"I liked the progression of literary works from Job to Anne Rice because we were able to see the formation of the concepts."

"Stereotypes were abolished, new perceptions formed, all due to the effects of this class and Professor Spumante."

"I really liked the class and the way it made me question a lot of my beliefs and ideas that I had taken for granted or never really thought about before."

"I learned more about biblical/satanic history than in my 14 years of religious education classes. It was stimulating and interesting. One of the better, or best, professors I've had in my four years."

"Professor Spumante was always very energetic in his teaching style, often passing that enthusiasm onto his students. He also made himself very available to his students and is always responsive to our ideas and thoughts."

"He makes us think and question and become passionately involved in what we are reading."

"This was the type of class I envisioned before coming to college, one that encourages discussion and thinking, and I'm glad I was finally able to find one before graduating."

English 221W "English Renaissance Literature" (Spring 2003)
24 Students
Overall Mean: 9.6 of 10
Overall Median: 10.0

Student Comments

"Professor Spumante was able to take some of the most difficult literature there is and not only present it in an understandable and interesting manner, but to relate it to all sorts of things the class could understand."

"I have to say that this class is the first one I've really learned anything from in terms of writing ability. I credit that to Professor Spumante and the simple fact that he wouldn't give an A to anyone who could write a sentence. Although I don't think that my grade will be as high as in other courses, I am incredibly pleased with the knowledge that I gained in this course."

"This course was constructed in such a way as to encourage considerate and considerable thought about the texts, as well as the contexts in which they were written. Paper criticisms were detailed and thoughtful."

"His enthusiasm exceeds that of any professor I've had."

"Sitting in a circle and discussing important issues, we read great literature, and Dr. Spumante showed a lot of passion for what he was teaching. I learned more in this class than I have in any other class in the 4 years I've been in college."

English 230.03 "Shakespeare" (Spring 2003)
200 Students
Overall Mean: 9.5 out of 10
Overall Median: 10.0

Student Comments

"Relating Shakespeare's works to film and modern media was very helpful in bringing Shakespeare out of the 1600s."

"It was the first time I ever felt that Shakespeare had anything to do with [modern] life."

"I enjoyed this class a lot because of the critical thinking aspect. I was forced to analyze the plays through different perspectives."

"This was one of the best classes I've had at UConn, even though I never really had an interest in Shakespeare. I learned a lot, came to every class, and was sincerely inspired."

"Professor Spumante's breadth of knowledge and passion for his subject has given me a new interest in classical learning."

English 230.05 "Shakespeare" (Spring 2002)
70 Students
Overall Mean: 8.6 of 10
Overall Mean: 9.0

"Superb course. Shakespeare now seems accessible to me and extremely modern!"

"His vast knowledge of both subject matter and background and the use of multimedia to exemplify points [was the most positive aspect of the class]. The stimulation of interest is wonderful and it is fun and exciting to come to class."

"Professor Spumante was extremely passionate and excited about teaching this course. I was dreading taking a class in Shakespeare, but he made the class very enjoyable."

"As a gay student, I valued Dr. Spumante's all-inclusiveness of queer issues. . . . Both stimulated my interest as well as comforted me by representing diversity in ideas and history."

"Professor Spumante allowed us to freely explore different avenues of thought both in our class discussions and in our written analyses of the plays. His enthusiasm . . . provided an excellent atmosphere for learning and enjoying Shakespeare."

English 230.05 / Spring 2004
RECONSTRUCTING SHAKESPEARE
TR: 2:00–3:15 in 434 CLAS

Professor: Georgio Spumante
Office: 132 CLAS
Phone: 486–4762

Office Hours: Wednesday: 1:00–2:30;
Thursday: 12:10–1:45
E-mail: spumante@email.edu

Course Description: In this introductory course, we will closely examine eight of Shakespeare's plays and their reconstructions in a variety of interdisciplinary media ranging from music to film. The recent explosion of Shakespeare in film, music, and the visual arts can be understood as the culmination of a much longer history of reconstructions dating back as far as the seventeenth century. How have such works figured in the construction of the almost mythological literary figure we recognize today as Shakespeare? How have they served to ensure or perhaps undermine his lasting place at the center of the English literary canon? What are the cultural-political costs and/or benefits of such reconstructions? Our main focus will be on the plays themselves, of course, but it is my hope that these media exercises will help us to think—in exciting and immediately relevant ways—about Shakespeare's place in modern American culture.

Texts: (Please purchase the Signet editions ordered for this class)

The Taming of the Shrew	*Twelfth Night*
A Midsummer Night's Dream	*Hamlet*
Romeo & Juliet	*Othello*
Richard III	*King Lear*

&

One $5 to $15 ticket to see the Connecticut Repertory Theater's Production of *A Midsummer Night's Dream* (April 15–17, April 21–25). Call ASAP for tix at 486–4226, and be sure to ask for "Student Discount."

Requirements:

1. Attendance, preparation, professionalism, and participation	15%
2. 8–10 unannounced reading quizzes	25%
3. 4-page midterm paper	20%
4. 3-page review of theater production	20%
5. Comprehensive final in–class examination	20%

Class schedule: (film schedule below)

T Jan. 20	Introduction to course; Trailer for Kenneth Branagh's 1996 *Hamlet*
R Jan. 22	The Renaissance and Shakespeare's Life
T Jan 27	*The Taming of the Shrew* ("Induction" and Acts 1–2)
R Jan 29	*The Taming of the Shrew* (Acts 3–5)
T Feb. 3	*The Taming of the Shrew*; Film: Sam Taylor's 1929 Film Version and the Concept of Textual Indeterminacy
R Feb. 5	*Richard III* (Acts 1–2)
T Feb. 10	*Richard III* (Acts 3–5)
R Feb.12	*Richard III*; Film: TBA: Metatheatricality
T Feb. 17	*Romeo and Juliet* (Acts 1–2)
R Feb. 19	*Romeo and Juliet* (Acts 3–5)
T Feb. 24	*Romeo and Juliet*; Film and Music: *West Side Story*, Radiohead, and Luhrmann's Adaptation
R Feb 26	*A Midsummer Night's Dream* (Acts 1–2)

T March 2	*A Midsummer Night's Dream* (Acts 3–5)
R March 4	*A Midsummer Night's Dream*; Music and Film: Mendelssohn's "Overture" and Performative Interpretation
T March 9	No Classes: Spring Break
R March 11	No Classes: Spring Break
T March 16	*Twelfth Night* (Acts 1–2)
R March 18	*Twelfth Night* (Acts 3–5)
T March 23	*Twelfth Night*; Media Exercise TBA; **Midterm Exam Due**
R March 25	*Hamlet* (Acts 1–2)
T March 30	*Hamlet* (Acts 3–5)
R April 1	*Hamlet*
T April 6	*Hamlet*; Some "paintings of Ophelia"; *Hamlet* in Pop Culture
R April 8	*Othello* (Acts 1–2)
T April 13	*Othello* (Acts 3–5)
R April 15	*Othello*; Film: Tim Blake Nelson's "*O*"
T April 20	*King Lear* (Acts 1–2)
R April 22	*King Lear* (Acts 3–5)
T April 27	*King Lear*; **Reviews Due (with ticket stapled to Paper)**
R April 29	*King Lear*; Jocelyn Moorhouse's version of Jane Smiley's *A Thousand Acres* and Revisionism

Final Exam time TBA

SAMPLE DISSERTATION PROSPECTUS

Josh Irving
University of Graduate Work/Medieval Studies Program
May 12, 2004

Conditioning the Soul: Spiritual Athleticism in Medieval English Theology and Literature

I. Purpose, Importance, and Novelty of the Study

In this study I will explore the importance of spiritual athleticism for understanding the religious and literary worlds of medieval England. I will consider in some detail the birth and development of this phenomenon—wherein theologians considered Christian ascetics, martyrs, saints, and virgins to be Athletes of Christ or Athletes of God—from its biblical roots through its manifestations in patristic texts and early hagiography. The bulk of my study, however, will focus on England between the years ca. 700 C.E. and ca. 1485 C.E. It is my contention that in this roughly eight hundred year span, the idea of spiritual athleticism had a profound effect upon and was profoundly affected by English history, theology, and literature. By examining contemporary hagiographic works, spiritual texts, mystical writings, and imaginative poetry and prose, I intend to demonstrate both the significance of spiritual athleticism for Christian thought in the English Middle Ages and also the progression of the spiritual athlete from theological *topos* in early medieval English texts to metaphorical construct in late medieval English literature.

The idea of spiritual athleticism itself, although a very important element of early Christian thought and, indeed, frequently found in ancient and medieval texts, has been all but ignored in modern scholarship. Most scholars of history, religion, and literature choose to examine another prevalent figure of the time—the Soldier of Christ (*miles Christi*)—in great detail,[1] relegating spiritual athleticism to a short paragraph or even just a footnote in their work, as if the spiritual athlete were an afterthought or somehow secondary to the spiritual warrior rather than one part of the same continuum.[2] Furthermore, no scholar to date has discussed even tangentially the effect of this athletic *topos* on the Christianity of medieval England or its manifestation in English religious and literary texts of this era. It is the intention of this study to explain and correct this oversight and to examine spiritual athleticism in fresh, focused ways in the hopes of unlocking its far-reaching significance for medieval English thought.

When writers of the late antique/early medieval period referred to particularly devout Christians as athletes, they were using the writings of St. Paul as a guide. . . . (**several pages cut**)

II. Methodology

The roots of spiritual athleticism that I have outlined above (and that I will develop in much more detail in the dissertation) provide the groundwork for my examination of the *topos* in medieval English religion and literature. I will be more specific about the development of my argument in the section of this prospectus in which I break down the subject matter of each individual chapter, but the general direction of my study will be to begin with the Anglo-Saxon appropriation of spiritual athleticism in religious works. After the Anglo-Saxon period, though, references to religious men as spiritual athletes seem to die away. The *topos* remains, however, in English religious writings by women, for women, or both. Thus, my study will next move into an analysis of Anglo-Norman and Early Middle English spiritual texts that concern women. Finally, I will move into what I see as the transformed and popular use of the spiritual athlete as a trope in later medieval literature.

I will also engage several critical discourses in my project. First of all, the project will involve itself with the language and methodology of social history, particularly that school of social history that has recently been gaining prominence: the history of sport. I will use the methodology of the sports historian in two different ways. In the first place, I will examine the cultural importance of athletics as a basis for my study of the Pauline tradition. To do this I will enter into the discussions of Johan Huizinga and Clifford Geertz, among others. Huizinga and Geertz both offer cultural/anthropological analyses of the origins and meanings of play and sport. Huizinga, in *Homo Ludens*, studies play in a very broad context, focusing on the impact of a common play-element found in all cultures. Geertz, on the other hand, concentrates specifically on the ritual-contest of cockfighting in Balinese society in his seminal article "Deep Play: Notes on a Balinese Cockfight." Geertz's article seems in some ways to be a response to the widely-influential work of Huizinga. By focusing on one culture, Geertz seeks to eliminate the sweeping generalizations of a work that is as grand in scope as Huizinga's. He succeeds to the extent that he shows the cockfighting contest to have real-life implications (beyond the play-world that Huizinga privileges) carried out in "play form" (325).

I have chosen these two texts initially because they represent the two extreme ends of the spectrum of cultural studies of sport. As disparate as they are, the idea of contest that they discuss will be important for my own research. My study will employ a combination of their approaches as I determine the cultural importance of contest for Christian thought. I will examine the phenomenon widely at first, as does Huizinga, so that I may get a sense of the broader understanding of the spiritual athlete in early Christianity, but then I will look very specifically at medieval England to gauge how this construct helped to define the spirituality of the English Middle Ages.

The second way in which I will use the methodology of the sports historian is to explore primary texts for information about the practice and culture of medieval sports. . . . **(several pages cut)**

III. Chapter Outline

Introduction

In short, the introduction to my dissertation will serve three purposes. I will first discuss the idea of spiritual athleticism in general with an analysis of the scriptural precedence for the phenomenon in much the same way that I have presented the material in the first section of this prospectus. I will expand what I have presented thus far, however, by including a more detailed approach to the story of Jacob and how this biblical tradition works with the Pauline Epistles to create what we know of as spiritual athleticism, along with some possible implications that the idea of the athlete can have for our understanding of the development of Christianity. I will also provide the background for the *topos* found in early exegetical works and early hagiography, such as the *Ad Martyras* of Tertullian, the commentaries on Paul by Rabanus Maurus and Ambrosiaster, and the *Vita* of Antony by Athanasius. After I establish this I will also clarify the distinctions between the spiritual warrior and the spiritual athlete that many scholars gloss over entirely. It seems to me that it is the idea of discipline that separates the athletic tradition from the military tradition. Whereas battle is predominantly concerned with victory and active means to achieve such victory, athletics concerns the conditioning of the body as a means of preparation. It follows logically that the *miles Christi* aggressively pursues war against demons, sin, and temptation, while the *athleta Christi*, as I will show in more detail below and in the bulk of the work itself is very rooted in the preparation of his or her body before the contest with temptation. The second purpose of my introduction will be to expand upon the methodological considerations that I have presented above.

OTHER CHAPTER SUMMARIES FOLLOW (NOT INCLUDED HERE)

IV. Availability of Resources and Time Schedule

Homer Babbidge library has many of the materials that I will need to complete this project, but I will have to use the Inter-Library Loan system to retrieve some of the more obscure texts. I also anticipate having to spend a summer or a semester (probably in 2005) in England investigating manuscripts in the British Library and other locations. This will be particularly important for chapter two, in which I deal with the nature of the word *cempa(n)*, because I will need to provide manuscript evidence to prove the slippery point about the term's use to mean "athlete."

Though my project covers a large amount of material, I think that it is reasonable to complete it within two years. If we include the introduction, then the dissertation is effectively divided into five sections. This would allow me to spend nearly a semester (plus two summers) on each section; this certainly seems ample time to see the project through to its completion. I do not anticipate that my research in England will increase the time needed to finish my project, because I will be writing parts of the dissertation while I am researching others.

BIBLIOGRAPHY (FOLLOWS HERE)

SAMPLE COVER LETTER FOR CONFERENCE PAPER PROPOSAL

University Letterhead

May 15, 2001

Professor Sharon Reinholder
Department of English
University of Maryland
College Park, MD 20742

Dear Professor Reinholder:

Please consider the enclosed abstract, " 'Heroic games' and 'Idle pastimes': Milton's Ambivalent *Uses* of Sport and Recreation," for presentation at the Seventh International Milton Symposium in June 2002. The paper considers Milton's *seemingly* contradictory attitude toward sports and recreations in light of seventeenth-century distinctions between "profitless" and "functional" physical activities.

Would you please use the following address for any mail after June 30: 27 Holly Drive / Storryville, MA 09861. I am currently in the process of moving from Old University to New University and do not yet have a new office mailing address. My current e-mail address is milton@edu. Sorry for the confusion.

Thank you for your time and consideration.

Sincerely,

Sarah Inkling
Assistant Professor
Department of English
New University

Sample Conference Abstract (from Previously Written Piece)

"Heroic games" and "Idle pastimes": Milton's Ambivalent *Uses* of Sport and Recreation

According to Christopher Durston, one of the "prime targets of puritan reformers" during the Interregnum was the traditional festive calendar, along with the traditional sports, games, and pastimes of the English people. The battles waged between royalists and parliamentarians in the 1630s—over the legality and morality of the sabbath-day sports endorsed by Charles I in the 1633 reissuance of his father's *Book of Sports*—ended abruptly with the victory of parliament in the civil war. The official burnings of the *Book of Sports* in 1643 were followed by legislation declaring that "no person or persons shall hereafter upon the Lord's-day use, exercise, keep, maintain, or be present at any wrestlings, Shooting, Bowling, Ringing of Bells for Pleasure or Pastime, Masque, Wake, . . . Games, Sport or Pastime whatsoever." Subsequent ordinances of 1653, 1654, and 1657 extended the legislation to include such sports as animal baitings, cockfights, and horse races. Though, in recent years, most cultural historians have supported Kenneth Parker's argument that the "gap between regulation and enforcement [of such policies] remained quite wide" throughout the Interregnum, the government's Draconian measures did mark at least the *official* death of sport in England between 1642 and the Restoration. The raising of Maypoles all over England, however, signified the triumphant return of the King in May 1660. Along with the Maypoles returned the old festivals of the Christian calendar and "official tolerance" of the old pastimes.

Based on this somewhat overly teleological—though essentially accurate—history, scholars have too readily associated a contemporary "pro-sport" polemic with royalists and an "anti-sport" polemic with parliamentarians and the godly. A thorough investigation of Milton's multiple uses of sporting metaphors and imagery, however, reveals the author's ambivalent attitude toward sports and recreations, an attitude quite typical of both radicals and conservatives in the early modern period. Though Milton advocates sports such as wrestling and fencing in *Of Education* (Part III), sports that keep soldiers "healthy, nimble, and strong," he condemns the Philistine "wrestlers, riders, [and] runners" of *Samson Agonistes* (1324). In *Paradise Lost*, both the virtuous and fallen angels participate in athletic games. In Book Two, Satan's legions participate in various sports while their lord is away: "Part on the plain, or in the air sublime / Upon the wing, or in swift race contend, / As at the Olympian games or Pythian fields" (528–30). The virtuous angels play at similar sports: "About [Gabriel] . . . exercised heroic games / the unarmed youth of heaven . . ." (4.551–2).

In the proposed paper, I will consider Milton's *seemingly* contradictory attitude toward sports and recreations in light of seventeenth-century distinctions between "profitless" and "functional" physical activities. Detailed consideration of the political/religious factors influencing contemporary attitudes toward sport accomplishes three goals in particular: first, it helps us to recover the allegorical meanings underlying Milton's multiple references to sport; next, it deepens our understanding of sport's complex centrality in the early modern imagination; finally, it challenges our current understanding of the seventeenth-century controversy surrounding sports, recreations, and mirth.

Sample Conference Abstract Generated for Conference

Looking for Shakespeare in *The Animated Tales* Film Series

Released over a two-year period beginning in 1992, *Shakespeare: The Animated Tales* featured 13 short films based on Leon Garfield's award winning children's book series, *Shakespeare Stories.* Jointly produced by S4C (Channel Four Wales) and Moscow's Soyuzmultfilm Studios, the project necessitated more than 600 animators working with glass, cell, and clay animation techniques. From the beginning, the goals of this ambitious project were clearly defined, as a promotional advertisement from 1992 reveals:

> Using a medium that's universally understood and enjoyed, Shakespeare's valuable cultural heritage is introduced in an accessible, exciting form to enthrall, encourage and educate. Skillfully condensed to half an hour each, The Animated Tales are the ideal length for the targeted audience of 10–15 year olds and will in addition have wide family appeal.

Through the "accessible" and "universal" language of animation, the producers of *The Animated Tales* will pass onto another generation the valuable cultural institution that is Shakespeare.

I am less interested in critiquing such a project (cultural materialism, after all, has practically rendered such critical moves redundant) than in evaluating the *actual* "cultural heritage" passed on to our 10–15 year olds through the *Tales.* I should like to argue that Shakespeare, as the mythological conveyor of Western values, is paradoxically written out of the *Tales* in the translation from play text to animated film text. Because individual films have a maximum running time of about 25 minutes, each director is forced to make drastic reductions of the play text. As a brief description of Nikolai Serebryakov's *Macbeth* demonstrates, the final product usually consists of plot summary and a few famous lines from each play:

> Macbeth! The very word conjures up a world of darkness and blood. The most terrifying of Shakespeare's plays, evil rises from it like a black fog and overwhelms all light of day. Set in Ancient Scotland, it is a tale of murder, madness and the huge ambition of a husband and wife, that finally destroys them. Macbeth and his wife begin by murdering the king and end by murdering their own souls; "I have supped full of horrors," groans Macbeth.

The brief advertisement accurately describes what we eventually find in Serebryakov's film: an exciting story and a few famous quotations.

Ironically, in an attempt to pass on the Shakespearean cultural heritage through an accessible medium, the makers of the *Animated Tales* are forced to ignore *character*, which has been historically regarded as the most important marker of Shakespeare's value, universality, and literary superiority. What does it mean to reduce *Macbeth* to a narrative originally written by Raphael Holinshed? How does "Shakespeare" signify when *A Winter's Tale* is made indistinguishable from its source, Robert Greene's *Pandosto*? In essence, *The Animated Tales* reconstructs Shakespeare, not as the conveyor of eternal values or the inventor of Western interiority—as the advertisement would imply—but as a great storyteller. But even Garfield's original title, *Shakespeare Stories*, is somewhat misleading for the obvious reason that few of the stories told by Shakespeare were really his own. Based on these conclusions, I propose an in-depth exploration of the manner in which the *Animated Tales*' reconstruction of Shakespeare as storyteller transforms, even as it perpetuates, the Shakespearean cultural heritage.

Sample Submission Cover Letter

University Letterhead

June 12, 2002

The Chaucer Journal
Robert Worler, Jr.
English Department
117 Burrowes Building
The Publishing University
University Town, FLA 16802

Dear Professor Worler:

Please consider my manuscript—"Athletic and Discursive Competition in Fragment I of the *Canterbury Tales*"—for publication in *The Chaucer Journal*. I have enclosed two copies as requested.

Should you need to contact me, I can be reached by phone at (860) 429–9106 or by e-mail at aspiring@medievalist.com. My address is listed below and on the first page of the manuscript. Thank you for your time and consideration.

Sincerely,

Joan Baez
Assistant Professor of English

SAMPLE COVER LETTER FOR FINAL SUBMISSION

University Letterhead

November 29, 2002

The Chaucer Journal
Professor Robert Worler, Jr.
English Department
117 Burrowes Building
The Publishing University
University Town, FLA 16802

Dear Professor Worler:

Please find enclosed the following materials: 1) one hardcopy of my revised essay, "Historicizing 'Wrastlynge' in the *Miller's Tale*," which was accepted for publication on September 19; 2) a disk copy of the essay saved in Microsoft Word 97; 3) a signed copy of the Pennsylvania State University Press publication agreement; 4) and an attached description of my revisions.

Should you need to contact me, I can be reached by phone at (860) 429–9106 or by e-mail at aspiring@medievalist.com. My mailing address is listed below and on the first page of the manuscript. Please note that I am happy to receive e-mail attachments. Thank you for your time and cooperation.

Sincerely Yours,

Joan Baez
Assistant Professor of English

SAMPLE LIST OF ITEMIZED REVISIONS

Re: Explanation of Revisions for Joan Baez, "Historicizing 'Wrastlynge' in the *Miller's Tale*."

First of all, I would like to thank the anonymous readers of *The Chaucer Journal* for their perceptive and helpful comments regarding my essay. I have taken all of them into consideration, and I believe that the article is stronger as a result of the revisions, which include the following:

1. The most serious revision involves my decision to cut almost all of the comparison of the *Knight's Tale* and the *Miller's Tale*, which was recommended both in the longer reader's report and in the editors' cover letter. The original 43 page article has been tightened considerably and is now about 28 pages.
2. Based on recommendations from both readers and from the editors, I have changed the title to reflect more accurately the purpose and scope of the essay. As the new title would suggest, the revised essay focuses more fully on the historical context of Chaucer's tale than on the competition between the Miller and the Knight.
3. All of the minor changes recommended in the shorter report have been addressed. The fourth point—that the "woman-as-prize" issue warrants further discussion—is no longer relevant in light of the revisions that I've completed.
4. The longer reader's report was extremely valuable because it pointed out several mistakes in my original organizational approach and in my use of certain poorly defined terms. My essay is undeniably stronger because of this reader's generous attention to detail. The comments about the flawed rhetorical question on page 12 and my misuse of the terms, "classless" and "collapsible," for instance, have helped me to revise. Many of the comments (almost all of the comments on the final page of the report, for example) no longer apply because of my decision to cut the section comparing the tales. I disagree with several of the reader's suggestions, however, and I would like to explain here the reasons for my disagreement:
   (a) The reader's description of "the fallacy of the undistributed middle term" (page 1 of the report) claims that I am equating wrestling matches and tournaments as though they are the same thing. Actually, I wish to do no such thing, and I have revised my language in every passage that even hints at this sort of logical (or illogical) move. Rather I am pointing out that the Miller and the Knight share one feature in common: their skill in one-on-one combat. After *differentiating* such activities as tournaments and common wrestling matches, I explain that tournaments—like many knightly activities—often included wrestling matches, in order to reinforce my point that Chaucer's coevals would have associated wrestling with members of the upper as well as the lower orders. I then discuss the implications of this fact for the reader's understanding of the Miller's attempt to "quite" the Knight's tale. My argument that contests-within-the-tales can operate as analogues for and commentaries on the storytelling contest itself seems to me perfectly logical, and I believe that the argument is supported by the evidence that I provide.
   (b) Because I do not wish to argue that wrestling matches are synonymous with tournaments, I have decided to avoid discussing the Medieval tournament at greater length (though I have deepened somewhat my

recognition of the tournament literature). The subject of the article is the Medieval wrestling match, and I am only interested in the tournament insofar as it helps me to historicize wrestling and, more specifically, the role of the knight as a wrestler.

(c) The reference to the Lambdin's "non-sensical" description of the historical figure of the miller on page 14 has caused me to cut the clause specifying that the millers lacked "an identifiable stature." I agree with the reader that the Lambdins' terms are slightly off. I have retained much of the original quotation, however, which simply summarizes recent critical evidence (including that offered by Lindahl) that millers were partly defined by their exceptional social flexibility.

(d) Finally, in response to the reader's suggestion that I cut all references to Terry Jones' book, I offer instead a very clear disclaimer in this section specifying my *qualified* agreement with Jones' major claim. The subsequent characterization of the Knight and the Miller as figures occupying nebulous positions in an "increasingly blurry social hierarchy" is backed up by Lindahl, among others, and the extensive historical details regarding Medieval wrestlers and knights. My point is that although the CT is obviously a fiction, it is best understood within its specific historical context, and this context includes the fact that knights were gradually being replaced in the period by mercenaries.

5. The longer reader's report suggested that I cut the material on the previous literature (Pindar, Homer, etc.) that conflated discursive and athletic contests. As a result, I have decided to cut the entire section.

SAMPLE COVER LETTERS FOR BOOK PROSPECTUSES

University Letterhead

January 3, 2002

Ms. Kathleen Avanto
Assistant Acquisitions Editor, Humanities
The Greatest University Press
2715 North by Northwest Street
Baltimore, Maryland 21218

Dear Ms. Avanto:

I am writing to inquire whether you would be interested in reviewing my book manuscript, "Unlawful Recreations: Sport, Politics, and Literature in Early Modern England," for publication by Greatest University Press. My interdisciplinary work on Early Modern culture and literature ties in nicely with recent GUP monographs on the cultural history of the period, including Raphael Donatello, *Charismatic Athletes*, Paula Johnson, *Spectacular Sports*, and Joan Hales, *War and Sport in Ancient Greece*.

Despite recent critical interest in nearly every aspect of Early Modern English popular culture, scholars have ignored sport, exercise, and athletics. This neglect is puzzling since sport occupied an integral position—both literal and metaphorical—in politics, medicine, military science, and art. To the degree that Early Modern scholars have studied "sport" at all, they have tended to conflate athletics and mirthful, disorderly activities such as drinking and gambling. In contrast, my book demonstrates that sport was central to Early Modern conceptions of order, health, and nobility, and it shows how major writers like Shakespeare and Milton used contemporary controversies about sport as a vehicle for social commentary and protest.

The critical response to this project—from colleagues in English, History, and Comparative Literature—has been enthusiastic. Several preliminary ideas are developed in essays published, or accepted for publication, in *SEL*, *Renaissance Quarterly*, and *Prose Studies*.

The completed manuscript, including the bibliography, is approximately 320-pages typescript (78,076 words, excluding bibliography) and requires no special design attention.

I hope that you will be interested in reviewing my book manuscript. Enclosed you will find a brief prospectus, chapter outline, introduction, sample chapter, and vita. I can be reached by phone at my office (860 486–4762) or home (860 429–9106) and by e-mail (semenza@uconn.edu). Thank you for your time and consideration. I look forward to hearing from you.

Yours Sincerely,

Gregory M. Colón Semenza
Assistant Professor of English

University Letterhead

January 20, 2004

Farideh Koohi-Kamali, Editor
Palgrave Macmillan
175 Fifth Avenue, Room 203 (WB)
New York, NY 10010

Dear Ms. Koohi-Kamali:

We are writing to inquire whether you would be interested in reviewing our book project, *Milton in Popular Culture*, for publication by Palgrave Macmillan. We feel that Palgrave is perhaps the single most appropriate press for our book, mainly because of your serious commitment to Renaissance studies, including Milton scholarship (Jordan, Wynne-Davis, and Maley), and your recent movement to the forefront of the Shakespeare and popular culture field (marked by the publication of important works by Deborah Cartmell and Richard Burt). A discussion of the project with Melissa Nosal, at the recent MLA convention, confirmed our sense that Palgrave would be ideal for this project.

While some two dozen books on Shakespeare and modern popular culture (especially film) have appeared in the past two decades, no comparable study of John Milton exists. The provocative and widespread appropriations of or allusions to Milton in film, television, fantasy literature, popular biography, music, newspapers, and the web have been all but ignored by the academic community. Addressing this gap, our volume brings together both younger and more senior scholars—specialists in the Renaissance or in popular culture/film—who explore how Milton, canonical writer *par excellence*, both influences and is influenced by a variety of forms of popular culture. Senior Miltonist and public intellectual Stanley Fish has agreed to contribute an Afterword for the volume.

Ranging from classic film (*Sabrina, The Lady Eve, Bride of Frankenstein*), to contemporary film (*Last Temptation of Christ*, *Dogma*, *The Devil's Advocate*), to television (*Star Trek*), popular literature (Philip Pullman, graphic novels), social activism (Milton Society for the Blind), the web, and the news, the rich and varied essays in our volume show how Milton plays a crucial role in popular culture and, in turn, how popular culture adapts and transforms Milton. Far from threatening the Miltonic legacy, such appropriations seek out Milton as a mode of legitimacy or as means of exploring issues of liberty, justice, good and evil, free will, gender roles, and republicanism. Such texts might allude to Milton, appropriate Miltonic language in surprising or subversive contexts, or evoke, grapple with, or contest Miltonic theodicy or gender hierarchy. In doing so, popular forms give new currency to Milton, making his works a vital, living part of contemporary culture.

Milton in Popular Culture is aimed at a broad range of scholars and teachers, including teachers of graduate and undergraduate Milton courses and of "Major British Authors" courses, as well as teachers of courses on popular culture. Milton courses are taught annually at colleges and universities across the English-speaking world. An inexpensive paperback detailing the uses of Milton in film, popular literature, social activism, and the web would be a unique and invaluable resource for those teachers, as they seek to make a seemingly dogmatic and difficult writer appealing and accessible to their students. A high-quality book on the subject of Milton and popular culture would undoubtedly sell internationally as well as in the United States.

This book project, on which we have been working for about a year, is at the advanced proposal stage. From a process of selection after a call for papers, we have now accepted or commissioned twenty essays of no more than 6,000 words. The completed manuscript will be approximately 125,000 words. Given the importance of the visual in popular culture and the advantages of using stills to illustrate discussions of film and digitized adaptations, the volume will include an appropriate number of black and white illustrations, approximately 30 (although this number is flexible). The contributors have agreed to submit their completed essays no later than May 1, 2004; we anticipate being able to forward the completed manuscript to a press by the end of August 2004.

Critical response to *Milton in Popular Culture* thus far has been enthusiastic. The Milton Society of America chose our topic for one of its two sponsored sessions at the December 2003 MLA Convention in San Diego. As editors, we were able to select from a wide range of submissions on this topic, including queries from scholars who specialize in film, contemporary fiction, or popular culture and have not previously had a venue in which to discuss their discoveries about Miltonic appropriations in these media.

The two editors are both well published on Milton and on Renaissance topics more broadly. Knoppers has a monograph on Milton—*Historicizing Milton: Spectacle, Power, and Poetry in Restoration England* (U of Georgia, 1994)—and has published more than a dozen essays on Milton, in addition to another book and two edited volumes on Renaissance topics. Semenza has a monograph on *Sport, Politics, and Literature in Early Modern England* (U of Delaware, 2004) and has published essays on such subjects as Milton and Shakespeare and film.

Enclosed you will find a detailed description of *Milton in Popular Culture*, including scholarly context and aim; detailed overview; target market; main competing books; detailed synopsis and chapter headings.

We hope that you will be interested in reviewing this project. We can best be reached via e-mail at semenza@uconn.edu or llk6@psu.edu, or at the office addresses on our enclosed c.v.s. Thank you for your time and consideration; we look forward to hearing from you.

Sincerely,

Gregory M. Colón Semenza and Laura L. Knoppers

Sample Book Prospectuses

The Self in the Cell: Narrating the Victorian Prisoner

Sean Christopher, Made-Up University

My purpose in this study of Victorian prison novels is to examine the emergence of the solitary confinement penitentiary in England, the psychological traumas that solitary confinement inflicted upon inmates, and the ways in which these traumas came to shape social perceptions about and narrative representations of Victorian prisoners. For twenty years, writers like Mark Seltzer, D. A. Miller, and Jeremy Tambling have used Foucault's discussion of the Panopticon in *Discipline and Punish* to formulate conclusions about the relationship between the Victorian novel and the prison. In particular, they have seized upon the idea of surveillance to help explain recurrent "metaphorical" forms of confinement like policing, detection, public scrutiny and censure, and even omniscient narration. Though these studies have in many ways been provocative, they ignore a serious problem: no prisoner ever spent a single day in the Panopticon, nor probably in a prison much like it. Instead, someone who broke the law in nineteenth-century England could expect to be transported to Australia to work under nearly unendurable privations; held for weeks or even months in an overcrowded and filthy local jail while awaiting quarterly assizes; committed for debt and thrust into the Marshalsea or the Fleet Prison, where a greedy jailer would demand exorbitant fees for a squalid cell and a pittance of food; or driven, as happened all too often, utterly mad by solitary confinement, the very tool that early-Victorian authorities hoped would produce moral regeneration and reform. Though these various kinds of prisons emerge again and again in Victorian novels, Foucauldian analyses have left these novels about real (rather than metaphorical) prisons virtually untouched—likely because these novels and their prisons have, rather inconveniently, very little to do with the Panopticon.

The Self in the Cell returns to the real forms of Victorian prisons and examines the ways in which the prisons of nineteenth-century England ultimately depended upon a model of solitary confinement and autobiographical production with enormously important narrative implications. This is a crucial return, for Foucauldian analyses have been notoriously neglectful of Victorian prison history. They have also, in their preoccupation with omniscient narration, ignored the role of private, first-person narration both in the Victorian prison and in the novels that center upon it. During the nineteenth century, the prison evolved as an institution specially designed to inflict deep and lasting psychological transformations upon those it confined, and also to force prisoners to narrate their psychological trials. In *The Self in the Cell*, I argue that the prison's power to inflict and narrate private trauma during the nineteenth century provided the impetus and the model for increasingly interior accounts of the private self.

Chapter One establishes the context for my discussion of the Victorian prison novel by tracing the cultural history of the Victorian prison. Beginning with a discussion of the old jails of the eighteenth century, it examines the historical and ideological reasons for the emergence of a new form of penitentiary predicated upon solitary confinement and prisoner autobiography. In 1775, before England turned to reformative imprisonment as a penal option, criminal punishment was deliberately punitive and visible: executions, stocks, pillories, and even brandings served to identify and injure those who had broken the law. Imprisonment, whether at Newgate or in local jails, likewise permitted and even encouraged contact and commerce between inmates and the public. But by 1850, with the solitary confinement penitentiary firmly established as England's primary sentencing option, punishment had become a much more private endeavor, expressly intended to remake convicts psychologically somewhere beyond the reach of the public stare. Most convicts endured some form of the national disciplinary program established by the Prison Inspectorate in 1835, which called for separate confinement intended to inspire self-reflection, moral regeneration, and (often) self-narratives that prison authorities read, edited, and interpreted in order to ensure that they told the "truth" about the prisoner's guilt and the beneficent effects of the cell. Thousands of others were transported to Australia, which received English convicts until 1868 and contained its own particular strategies for demanding and subjugating prison narratives. Locking inmates away in solitude or shipping them halfway across the globe, robbing them of their power to tell an unfettered account of their confinement, Victorians recreated the cell as a profoundly private place with substantial psychological and narrative complications for the confined. The result, I argue in this chapter, was a Victorian society that mostly experienced its prisons discursively, and that came to rely upon the prison authorities' own accounts of inmates' private psychological trials.

Subsequent chapters discuss several works by Dickens, Charles Reade's *It is Never Too Late to Mend* (1856), Marcus Clarke's *His Natural Life* (1870), Charlotte Brontë's *Villette* (1853), and Wilkie Collins's *Armadale* (1866). The aim of these chapters is to show that Victorian novelists began very early to recognize the psychological traumas inflicted by the new model of confinement, and that their determination to write realistic accounts of confinement drove them to adopt narrative strategies that allowed them to account for psychological trauma and aberration. As these chapters also show, Victorian novelists drew these narrative strategies—sometimes deliberately and consciously—from the prison. Chapter Two examines two of Dickens's earliest prison accounts, *Pickwick Papers* (1837) and *American Notes* (1842), both of which profess to be "honest" accounts of confinement and show a genuine anxiety about how to account for prisons that operate upon the mind much more powerfully than they operate upon the body. Chapter Three addresses Reade's novel about the Birmingham Gaol scandal of 1853, showing that—despite Reade's determination to write a documentary (if sensational) account of events—he takes great pains to mimic the prison's own methods of demanding and controlling prison autobiography. Chapters Four and Five return to Dickens, taking up *Little Dorrit* (1857) and *A Tale of Two Cities* (1859) to show that both novels insist that the prison is a private space, available for narration *only* within a first-person prisoner narrative that is invariably manipulated and marred by the prison's narrative power. Chapter Six, on Clarke's *His Natural Life*, extends this argument to the prison in Australia, showing how deeply intertwined imprisonment and psychological narrative had become by the second half of the century, even in accounts of the prison that did not hinge upon solitary confinement. Chapter Seven addresses *Villette* and *Armadale* and points out that, though neither novel contains a "real" prison, Brontë and Collins both seem to conceive of solitude and confinement as necessary pre-conditions to the processes of psychological self-accounting and self-ordering in which the female narrators of both books engage. The chapter helps to complete the argument that during the Victorian period the development of interior methods of narration *generally* came to depend upon the imagery and ideology of the solitary cell. The final chapter, Chapter Eight, discusses the broad implications of this argument, partly by returning to Dickens's unfinished novel *The Mystery of Edwin Drood* (1870) and arguing that this last book may well have been working to illustrate the conclusion that self-accounting and psychological exposition in the Victorian novel were impossible without the imaginative presence of the cell.

By arguing this in the final chapter, and by examining the rules of reading, interpretation, and narrative power that characterize prison novels' accounts of the self in the cell, I conclude that the Victorian prison novel presages the rules for psychological storytelling that Freud articulated at the turn of the century. To the extent that this is true, prison novels belong not only to literary history, but also to a cultural history of ideas about subjectivity, psychology, and narration.

Clearly this book will appeal first and foremost to scholars of the Victorian novel, who will (I believe) raise their eyebrows and breathe a sigh of relief over this book's provocative and historically grounded alternative to Foucault on the matter of nineteenth-century imprisonment. An occasional article has attempted to treat the "historical" Victorian prison in light of one particular Victorian text. But no one, to my knowledge, has offered this kind of sustained analysis of the real cultural relationship between the Victorian prison and nineteenth-century narrative practice. *The Self in the Cell* offers a new way of reading the literary significance of the prison, and also of reading the development of increasingly interior narrative fiction.

There is also a much wider audience for this book, consisting of historians, Dickens enthusiasts, and those interested in imprisonment and the origins of modern psychology. In many ways, our culture has not outgrown the Victorians' fascination with penitentiaries and criminals—with confessions, secret horrors, prison scandals, and the private infliction of insanity. Recent movies like *Murder in the First*, *Dead Man Walking*, and *The Green Mile* bear that out, as do new books by Norvald Morris and Peter Brooks. Though *The Self in the Cell* is "about" a serious scholarly topic, then, it has the power to attract thoughtful non-scholars as well. Just as important, the manuscript is written throughout in language that is sophisticated but accessible to the array of audiences who may take up the book. Readers of *The Self in the Cell* will make important discoveries about Dickens, Charlotte Brontë, and other Victorian novelists, to be sure. But they will also, I believe, come to fuller appreciations of the prison's importance to the development of interior narrative, psychoanalytic practice, and the shape of the modern novel.

MILTON IN POPULAR CULTURE

Edited by Laura Lunger Knoppers and Gregory Colón Semenza

Project Description

While some two dozen, highly influential books have appeared in the past decade on the subject of Shakespeare in popular culture, no such volume on Milton exists. Such neglect is, on the surface, odd, since Milton—like Chaucer and Shakespeare—is an industry unto himself; every English department teaches at least one Milton course annually at the undergraduate and often at the graduate level. While books on Shakespeare in film and other forms of mass media have transformed our perceptions of England's most famous poet-playwright, having become an invaluable tool in our productions and teachings of his works, no similar resources are available to Milton teachers and scholars. In short, the considerable presence of Milton in popular culture has been all but completely ignored. To address this gap, *Milton in Popular Culture* brings together both younger and more senior Miltonists and popular culture scholars from America, Britain, Canada, and Europe.

Milton has appeared quite recently in a range of seemingly unlikely places: classic films such as James Whale's *Bride of Frankenstein* and modern ones such as John Landis' *National Lampoon's Animal House* or Neil Lomax's *The Devil's Advocate*; in television programming as familiar as the revered original *Star Trek* series; in popular literature as critically acclaimed as Phillip Pullman's award winning *His Dark Materials* Trilogy and as cutting edge as the graphic novel *Sandman*; and in a number of other outlets as diverse as politically motivated editorials in the *Wall Street Journal* and the *Manchester Guardian*, and popular educational websites that receive thousands of hits annually. In light of such appearances, what is Milton's function in popular culture? What difference does it make that Audrey Hepburn's character, Sabrina, is named after Milton's major heroine in *Comus* or that Al Pacino's Satan happens to be named "John Milton"? Are conservative columnists correct when they claim that George W. Bush is the socio-political offspring of Milton? How does Helen Keller's founding of the still thriving John Milton Society for the Blind appropriate the famously blind poet in pursuit of greater social justice? This book sets out to answer these questions and more, showing how "Milton" transforms and is transformed by popular cultural modes and discourses.

To Miltonists and scholars of popular culture alike, Milton and popular culture might seem like dramatically antithetical subjects. Milton scholars both in America and Britain have focused in the past twenty years on exploring Milton's work in his specific historical, political, and literary contexts. Influences on Milton have received far more attention than Milton's influence on later writers, which has been restricted to canonical figures of the eighteenth and nineteenth centuries. Popular culture of the twentieth century has been largely overlooked. Generally speaking, we believe there are two explanations for this critical neglect. First, unlike Shakespeare—whose plays were performed in the public theaters of Southwark and constituted the equivalent of Renaissance pop-entertainment—Milton's complex prose and poetry were geared toward only a "fit audience, though few." As a religious and political radical, Milton never could have foreseen the impact of his writings on mainstream theological and political thought, let alone his influence on such "debased" art forms as film, fantasy literature, comic books, or heavy metal music. Second, Milton—even more than Chaucer or Shakespeare—became a lightning rod in the canon wars that rocked and, in some cases, divided English departments in the 1980s and 90s. For some scholars, defending the elitist Milton tradition has become tantamount to defending western civilization itself. For their opponents, however, Milton seems like the most logical target in an ideological campaign to stamp out distinctions between so-called highbrow and lowbrow art forms. The long-standing myth of "Milton's bogey" has ensured that Milton would continue to represent everything the field of popular cultural studies is against.

Although some might feel, then, that popular culture threatens our literary heritage, we would argue that studying it more closely actually highlights the vitality and relevance of Milton for the twenty-first century. For students, critical analyses of Milton's influence on popular films, comic books, and rock music have the power to transform a poet they previously assumed to be abstract and inaccessible. In fact, a book such as *Milton in Popular Culture* facilitates the process of passing down to a younger generation a Milton legacy already adapting to modern technologies, art forms, and popular discourses. For scholars, on the other hand, such a book reaffirms the basis of historicist scholarship and of liberal education itself: that in order to understand the past and the assumptions that inform such understanding, we need to be actively tuned in to the present. As the invocations of Milton in the news after 9/11 would suggest, Milton continues to shape us and our ways of thinking in the 21st-century.

Audience

Milton in Popular Culture is aimed at four major target audiences: Miltonists, teachers of Renaissance literature, popular culture scholars, and graduate and upper-level undergraduate students. Because the book can be used in a variety of graduate and undergraduate level courses, we envision it being adopted widely. The most obvious target audience will be scholars who write on Milton and teach his works in annual undergraduate and graduate courses. Filling a unique niche, this book could not only be used as a teaching resource but would also be ideally suited as a required text for either upper-level undergraduate or graduate student audiences. A less obvious but potentially equally important audience would be the hundreds of professors and graduate assistants who teach English literature surveys every semester in America, Canada, Australia, and Britain. For such teachers, many of whom are not Miltonists but regularly teach *Paradise Lost*, our book would be an essential tool for making Milton exciting and accessible. Finally, most English departments offer courses in literature and film and other forms of popular culture. A book on Milton in popular culture is likely to be useful to the teachers of such courses.

Competing Books

Although *Milton in Popular Culture* would have no direct competition, it can be most closely compared to books on Shakespeare and popular culture or to Milton companions and teaching volumes. The most relevant previous publication is Douglas Lanier's *Shakespeare and Modern Popular Culture*

(Oxford, 2002). Along with such groundbreaking books as *Shakespeare, The Movie* (Routledge, 1997), *Shakespeare and the Moving Image* (Cambridge, 1994), and *Shakespeare After Mass Media* (Palgrave, 2002), Lanier's work demonstrates the sort of broad interest a study of a canonical figure and popular culture can generate. Unfortunately for Renaissance scholars, such studies have been limited only to one author: Shakespeare. Studies of Milton most comparable to this volume would be Thomas Corns, ed., *A Companion to Milton* (Blackwell, 2001), Richard Bradford's *The Complete Critical Guide to Milton* (Routledge 2001), and Peter Herman's (ed.) forthcoming *MLA Approaches to Teaching Milton*, all of which are aimed at a broad academic audience but none of which deal with popular culture.

Contents

Excluding the introduction, chapters of 5,000 words or less are organized into six basic categories, which emphasize the comprehensiveness of the volume.

Introduction: "Milton in Popular Culture."
Laura Lunger Knoppers and Greg Colón Semenza.

Part I: Milton in Fantasy Literature

1. Figuring Milton in C. S. Lewis's Fiction
   Sanford Schwartz
2. Steven Brust's *To Reign in Hell*: The Real Story
   Diana Trevino Benet
3. "Fighting since Time Began": Milton and Satan in Philip Pullman's *His Dark Materials*
   Stephen Burt
4. *His Dark Materials, Paradise Lost, and the Common Reader*
   Lauren Shohet

(Rest of Contents Follows)

SAMPLE JOB APPLICATION COVER LETTER

Search SR October 17, 2000

University of Connecticut
Department of English U-25
337 Mansfield Road
Storrs, CT 06269–1025

Dear Search Committee Members:

I wish to apply for the Assistant Professorship in Shakespeare/Renaissance literature, which you advertised in the online edition of the *MLA Job*

Information List. On December 4, I will defend my dissertation—*Unlawful Recreations: Sport, Literature, and Politics in Early Modern England*—to complete the requirements for my Ph.D. in Renaissance literature at the Pennsylvania State University. I will receive my degree on time in May of 2001 and believe I am well qualified for the position you describe.

My dissertation argues that sport mediated the poles of excess and control in the Early Modern imagination. Despite contemporary polemical and even modern scholarly representations of sport as disorderly or carnivalesque phenomena, sport also figured at the center of Early Modern conceptions of order. In fact, many of the greatest contemporary prose writers—Elyot, Ascham, and Mulcaster, among others—contended that sport was vital to the health of the nation. In the opening chapter, I demonstrate the influence of Galen and other authorities on Early Modern proponents of sport, who claimed that sport was physiologically beneficial to the practitioner, militarily beneficial to the commonwealth, and socially beneficial to the maintenance of the reigning class system. Over the next five chapters, I explore the manner in which major Renaissance authors—including Shakespeare, Jonson, Drayton, Walton, and Milton—and a host of minor ones employed sporting imagery, metaphor, and allegory to defend or critique the social order that sport was believed to uphold. In its broad coverage of works published between Elyot's *Governor* in 1531 and Milton's *Samson Agonistes* in 1671, the dissertation considers the significance of sport in texts ranging from prose conduct manuals and sabbatarian pamphlets to drama, lyric poetry, and pastoral. Much of my dissertation has already been accepted for publication. An early version of Chapter One, "The Legacy of the Anti-Sports Polemic in Early Modern Prose," will appear in a forthcoming volume of *Prose Studies*. Part of the final chapter, "*Samson Agonistes* and the Politics of Restoration Sport," has been tentatively accepted by *SEL*, and the second chapter, on Shakespeare's *Henry VI*, has been revised and resubmitted to *Renaissance Quarterly*. As I prepare the manuscript for book publication, I plan to add a chapter exploring how Early Modern gender relations were influenced by the developing science we now call "exercise physiology."

By graduation, I will have taught 20 sections of 6 different courses at Penn State, mainly in our highly regarded Rhetoric and Composition program. In freshman and upper-level composition courses, I have capitalized on the small-classroom environment to emphasize critical reasoning and argument through a student-centered, dialogic approach. Due to my success in the program, the department has granted me the freedom to design and implement several of my own courses within the existing system. These advanced rhetoric courses, such as "Relativism and Absolutism" and "The Problem of Historical Interpretation," have helped students to become more critical readers and more articulate contributors to extremely complex discourses.

As a teacher in the Literature program, I have worked to bring the same focus on critical pedagogy into courses that often enroll more than sixty students. In Introduction to Shakespeare, for example, I highlight the concept of "textual indeterminacy" as a means of encouraging students to be more active and confident readers. This approach helps them to deconstruct the intimidating myth of Shakespeare as the conveyor of an eternal body of knowledge and frees them to interpret each play we study in class. In British Literature Survey to 1798, two students are required at the beginning of each class to present opposing arguments on a controversial aspect of the text due that day, and their arguments form the foundation from which each class conversation emerges. When combined with small group work, detailed feedback on written work, and frequent office conferences, such methods help to create the sort of small-class atmosphere that allows students to be more active learners.

Last spring, I received Penn State's most prestigious award for graduate instructors—the Teaching Assistant Award for Outstanding Teaching—along with only four other graduate assistants in the university. Based on the courses I have taught, my average "Quality of Instructor" ranking is a 6.27 on a 7.0 scale, and my highest scores (6.83 in Shakespeare, 6.65 in Freshman Composition, and 6.55 in the Literature survey) have well exceeded departmental averages. Next semester, I will become the first graduate instructor at Penn State to teach a 400-level Shakespeare course. My serious commitment to teaching is also reflected in the various teaching-related projects I've undertaken, which include publishing a short piece in *Shakespeare and the Classroom*, chairing a conference session entitled "Teaching Shakespeare and Film," and participating in a teaching workshop at last year's Shakespeare Association Conference.

Beyond the dissertation and the classroom, I have pursued numerous other professional interests. In 1996, I earned a research grant from the Folger Shakespeare Library to participate in a seminar taught by the president of the Shakespeare Institute, Peter Holland, on the growing field of Shakespeare and film. More recently, I was awarded a Mellon Foundation grant to attend a four-week seminar focused on the question of traditional versus revisionist criticism in Early Modern studies. I have also taken on several leadership roles at the university. Last year, I was elected president of the English Graduate Organization, an honor that afforded me the opportunity to represent more than 120 colleagues. My experiences as president, and my role as the departmental delegate in the university's Graduate Student Association, have taught me the significant value of departmental service.

I would like to bring my commitment to service, research, and teaching to the Department of English at the University of Connecticut. Enclosed you will find my curriculum vitae, dissertation abstract, and a writing sample. I have requested that Penn State forward you a copy of my dossier. I can also supply a teaching portfolio at your request. Please contact me at (814) 235–1647 or gms149@psu.edu if you would like to arrange an interview at the MLA Convention in December. Thank you for your consideration.

Sincerely,

Gregory Colón Semenza

Enclosures

SAMPLE TWO-PAGE DISSERTATION ABSTRACT (NONTRADITIONAL)

Dissertation and Research Summary: Gregory M. Colón Semenza
Unlawful Recreations: Sport, Politics, and Literature in Early Modern England

Value of Research Specialization	In the Early Modern period, sport was believed to be a highly significant phenomenon: physiologically beneficial to the individual practitioner, vital to the preparedness of the military, and necessary to the maintenance of the traditional class hierarchy. Sport's significance in the period is perhaps best registered by its literal and metaphorical centrality in numerous works of literature, including Shakespeare's *Henry VI*, Walton's *Compleat Angler*, and Milton's *Samson Agonistes*, among others. By reconstructing a cultural history of sport and investigating representations of it in contemporary prose, poetry, and drama, I demonstrate its pivotal position in the interlocking spheres of Early Modern science, politics, and art.
Argument of Dissertation	Early Modern scholars have long assumed sport to be synonymous with holiday mirth, drinking, and other such carnivalesque activities. I argue instead that sport mediated the poles of excess and order in the Early Modern imagination. While Bakhtin's influential definition of carnival as a "temporary liberation from the established order" would appear to describe unruly sports such as wrestling, these sports were just as often praised for contributing to the physical, mental, and social stability of the English people. The unusual ability of sport to navigate the extremes of order and chaos, morality and sin, function and superfluity made it an extremely effective vehicle for social commentary in practically every written form available to the Early Modern author.
Contribution of Dissertation	My dissertation expands and changes our current understanding of sport in Early Modern culture and literature. In my research I • Recover a conception of sport as a functional phenomenon. • Challenge the view that sport was primarily associated with carnivalesque activities. • Establish the manner in which authors commented on or attempted to subvert the status quo through the use of sporting imagery, metaphor, and allegory.

- Clarify the need for a more thorough scholarly excavation of the many texts and contexts within which sport figures so prominently.

Teaching Applications

The dissertation's wide coverage of culture and literature—between the publication of Elyot's *Governour* (1531) and Milton's *Samson* (1671)—has prepared me to teach both highly specialized courses on Early Modern culture and much broader surveys. My treatment of several major writers (including Shakespeare and Milton) and genres (prose, poetry, drama) insures a range of expertise in the classroom. In rhetoric and composition courses, my dissertation can be used to encourage students to think more critically about the social significance of phenomena often perceived to be apolitical or trivial.

Relevance to Future Research

I deliberately chose a dissertation topic that would provide a good deal of new material and a range of publication options. We currently lack a comprehensive study of sport in the period. My dissertation, once turned into a book, will begin to fill this gap in our scholarship. I also plan to produce a scholarly edition of a little-known manuscript in the Folger holdings, entitled *The Compleat Swimmer*. Finally, I have begun to research the influence of what we now call "exercise physiology" on Early Modern gender relations.

Chapter Breakdown: *Unlawful Recreations: Sport, Politics, and Literature in Early Modern England*

Chapter 1

"The Legacy of the Anti-Sports Polemic" charts the sixteenth-century construction of sport as a disorderly phenomenon. Moving from an in-depth discussion of the traditionally functional role of sport in England—as described by humanist prose writers (Elyot, Ascham, Mulcaster, etc.)—the chapter considers how the convergence of various social, economic, and intellectual factors enhanced the persuasiveness of the godly, anti-sports polemic.

Chapter 2

"Sport, War, and Contest in Shakespeare's *Henry VI*" continues the first chapter's consideration of the factors that contributed to sport's demise, focusing on innovations in military science. Whereas sport had been justified since antiquity for preparing soldiers for war, its function waned with the military's gradual adaptation of firearms. As a result, sport was increasingly condemned as a superfluous phenomenon. In *Henry VI*, sport figures as a metaphor for war, also condemnable as a result of the shift from the politics of chivalric idealism to the "politics of reality." Throughout the trilogy, Shakespeare indicts warfare as mere sport for ambitious nobles.

Chapter 3 "The Literary Context of the *Book of Sports* Controversy" demonstrates the mutually constitutive relationship between literary and political commentaries on sport. Investigating closely several anti-court satires written in the 1610s—including *Eastward Ho* and *The Isle of Gulls*—the chapter elucidates the manner in which sport was used by dramatists to critique or defend the political policies of James I. The primary focus of the chapter, however, is on James's *Book of Sports* as a reaction to such dramatic commentaries. In short, James's defense of lawful pastimes is a deliberate attempt to counter his popular reputation as an unlawful king.

Chapter 4 "The Burden of the Present" explores a relatively unknown collection of poems entitled *Annalia Dubrensia*, which featured work by Jonson and Drayton, among others. The poems celebrated the Cotswold Games, an English version of the ancient Olympics. The origins of sport, like those of poetry, trace back to Pagan societies, of course. To the degree that poetic identity depended upon a successful imitation of a pre-Christian society, the godliness of the 1630s threatened poetry as well as sport. By collapsing poetic and athletic competition, the *Annalia* defends the lawfulness of poetry at the same time that it defends sports.

Chapter 5 The chapter argues that Walton's *Compleat Angler* is an indictment of the governmental proscription of communal recreations in the 1650s. The anti-sport legislation of the Interregnum is the culmination of the Early Modern reconstruction of sport as a disorderly phenomenon, the process that I begin to chart in the first chapter. In Walton's pastoral world, the orderliness of society and, to a degree, of Nature, revolves around the ability of sport to bring people together.

Chapter 6 "*Samson Agonistes* and the Politics of Restoration Sport," brings us full circle by considering the resurrection of sport by Charles II and, more specifically, Milton's treatment of sport as a phenomenon mediating the poles of functionality and superfluity during the Restoration. Milton expands the meaning of the term "sport," which only implied "jest" in the *Judges* version of Samson's trials, to include athletic games and pastimes; he also employs the epithet "Agonistes"—a term used originally to describe ancient athletic contests—to characterize his protagonist. The deliberate contrast between Samson's functional athleticism, which serves both God and nation, and the Philistines' riotous sport at the Temple of Dagon allegorizes the massive gulf between the defenders of the good old cause and the corrupt new regime.

Sample Dissertation Abstract (Traditional)

The Self in the Cell: Narrating the Victorian Prisoner
Sean Christopher

Scholars have long depended upon Michel Foucault's discussion of the Panopticon, with its surveillant model of penal discipline, to shape discussions of the novel and the prison. Laden with omniscient narrators, critiques of social power, and portrayals of dangerous and criminal classes, the Victorian novel has been closely allied with this vision. I argue, however, that the real Victorian prison was based upon a model of separate confinement, self-reflection, and self-narrated guilt that has far different implications. The Victorian novel's relationship to the prison is best discovered not in discussions of surveillance and social power but rather in an investigation of the guilty autobiographical discourse that the prison both shaped and required. My project analyzes the ways in which Charles Reade, Wilkie Collins, Marcus Clarke, Charlotte Brontë, and—of course—Charles Dickens each wrote novels reflecting and recreating this separate prison.

In this project I achieve two aims. First, by recovering a historical prison unacknowledged by Foucault, I revise our understanding of what may properly be called the literature of the prison. Many of the greatest Victorian prison novels—Dickens's *A Tale of Two Cities*, Marcus Clarke's *His Natural Life*, and Charles Reade's *It is Never Too Late to Mend* among them—have little in common with Foucault's vision. As a genre they reflect instead a complex system of punishment that includes separate confinement, transportation, debtors' prisons, and disciplinary strategies with explicit (rather than metaphoric) narrative concerns. Demonstrating the place that narrative—particularly autobiography—occupies within this historical prison, I offer new ways of understanding prison novels' recurrent phenomena, especially their constant return to a submerged, psychological, yet essential first-person account of imprisonment.

Second, my project illustrates that as a genre prison novels raise crucial questions of narrative authority, psychological exposition, and the private self. Because reforming the prisoner depended so heavily upon producing personal feelings of guilt—really a new understanding of one's past life and its meaning—the separate prison required prisoners to express their past wickedness in autobiographies. The prison chaplain would "read" and "interpret" these to discover their hidden psychological meanings, since the aberrant, imprisoned convict could not be trusted to tell a true story. Asking its agents to invent this "true" account of the psychological self in the cell, the prison authorized the entry of the private mind into fictional accounts that could claim to present the "truth" of the psychological self. Prison novels recreate this narrative dynamic, their authors finding in the prison's narrative logic a justification for making psychological exposition a part of the narrative truth of the prisoner. They also illustrate time and again that escaping the prison often means reclaiming the power to narrate one's life, one's guilt, and one's experience in the cell. Recreating a narrative world of imprisonment that places the convict's private truth within a story told by a reforming, narrating other, prison novels presage the rules for psychological storytelling that Freud articulated at the turn of the century.

SAMPLE CONTENTS PAGE FOR CUSTOMIZED SYLLABI PACKET

Jeff Master

Sample Course Syllabi
For
2001–2002

Potential Hiring University

*Also prepared to teach:
English 3301: English Literature of the Renaissance
English 3306: Drama of the English Renaissance
English 4304: Selected Plays of Shakespeare I
English 5304: Studies in Renaissance British Literature
English 5305: Studies in Shakespeare
English 5306: Studies in 17th-Century Literature: The School of Donne
English 5306: Survey of Seventeenth-Century Literature
English 5351: Shakespeare and Film

SAMPLE THANK YOU LETTER POST INTERVIEW

University Letterhead

January 2, 2001

Professor David Jones
Chair, Department of English
Southern University
Georgetown, GA 78727

Dear Professor Jones:

Thank you for arranging an interview with me at the MLA Convention. I enjoyed the opportunity to speak with you and your colleagues from Southern. I particularly appreciated the clarity and forthrightness of the information you provided regarding your selection process. Please convey my thanks to everyone who was so generous with their time in speaking with me.

Should you need to contact me, I can be reached by e-mail at fakeuniversity @edu or phone (814–235–1647).

Sincerely,

Jeff Master

Sample Job Acceptance Letter

University Letterhead

February 9, 2001

Professor Janet Jones
Head, Department of English
Hiring University
337 Mansfield Road, U-25
Wonkaland, NY 00982

Dear Janet,

I am pleased to accept your offer of the position as Assistant Professor of English. Enclosed you will find a signed copy of the contract.

As I mentioned when I was on campus, I've felt very confident from the beginning of this job search that Hiring University is a place where I can be productive and happy. Certainly, the congeniality, professionalism, and candor of the search committee helped to affirm this feeling. I also appreciated your directness during the campus interview.

I'm looking forward to joining a strong department. I will be visiting Wonkaland for a few days in early March to look for a place to live. Hopefully, we can get together then.

Sincerely,

Wendy Whiner

NOTES

INTRODUCTION

1. *The Chronicle of Higher Education*'s annual *Almanac* provides a complete list of the Carnegie Foundation's classification of nearly 4,000 universities at http://chronicle.com/stats/carnegie/.
2. On time to completion issues, see The Survey of Earned Doctorates (SED), a report of "Doctorate Recipients from United States Universities: Summary Report 2002," which is sponsored by the National Science Foundation, the National Institutes of Health, the U.S. Department of Education, the National Endowment for the Humanities, the U.S. Department of Agriculture, and the National Aeronautics and Space Administration. See pages 21–23. This useful survey is accessible online at http://www.norc.uchicago.edu/issues/sed-2002.pdf.
3. Cary Nelson and Stephen Watt interrogate abuses of the term, "Apprentices," in relation to graduate students in *Academic Keywords: A Devil's Dictionary for Higher Education* (New York and London: Routledge, 1999), 58–71.
4. See Scott Smallwood, "Doctor Dropout," *The Chronicle of Higher Education* 50.19 :(2004) A10.
5. See *Higher Education Under Fire: Politics, Economics, and the Crisis of the Humanities*, ed. Bérubé and Nelson (New York and London: Routledge, 1995), 20.
6. Reported in the *Chronicle*'s 2009–10 *Almanac* which can be located at http://chronicle.com/section/Almanac-of-Higher-Education/141/.
7. To survey a list of unionized programs and those in the process of unionizing, see the website of the Coalition of Graduate Employee Unions at http://www.cgeu.org/contacts.html.
8. *The Chronicle of Higher Education* online features several good resources on "Non Academic Careers for Ph.D.s" at http://chronicle.com/jobs/archive/nonacademic.htm.
9. See http://www.inthemedievalmiddle.com/2006/09/on-beginning-graduate-school.html.
10. See the wonderfully useful blog of Dr. Virago at http://quodshe.blogspot.com/2006/09/to-professionalize-or-not-to_18.html.
11. *OED*, s.v. "Advisor."
12. Ibid.

13. Ibid.
14. Lewis Z. Schlosser and Charles J. Gelso, "Measuring the Working Alliance in Advisor-Advisee Relationships in Graduate School" *Journal of Counseling Psychology* (2001): 157–67 (157).
15. *Journal of Counseling Psychology* (2003): 178–88.
16. Ibid., 183.
17. Ibid.
18. Ibid.
19. Ibid.
20. Ibid., 186.
21. Yaritza Ferrer de Valero, "Departmental Factors Affecting Time-to-Degree and Completion Rates of Doctoral Students at One Land-Grant Research Institution" *The Journal of Higher Education* (2001): 341-367 (345).
22. Ibid.
23. Ibid., 356.
24. Schlosser and Gelso, 164, 165.
25. Schlosser, Hill, Knox, and Moskovitz, 181.
26. On this matter, see Michael Burawoy, "Combat in the Dissertation Zone," *American Sociologist* (2005): 43–56.

CHAPTER ONE

1. Very few universities actually issue six-year contracts. While most issue three two-year contracts, allowing for an official review of a candidate every two years, some issue six one-year contracts.
2. All salary estimates are based upon statistics compiled by the American Association of University Professors and published annually in the *Chronicle's Almanac of Higher Education*. In determining average faculty salaries for humanities scholars specifically I have estimated conservatively that humanities scholars earn approximately 10% less than the averages earned by professors in all academic fields.
3. The proliferation in recent years of Distinguished and Chair professorships is at least partly a response to people making full at relatively early ages and having no more promotions to shoot for. Such positions also serve as mechanisms by which hiring schools are able to attract distinguished senior professors from other institutions.
4. The best report on contingent faculty in America is the AAUP's policy statement on "Contingent Appointments and the Academic Profession," which can be viewed online at http://www.aaup.org/statements/SpchState/contingent.htm.
5. For a good summary of what's being done about it, see Scott Smallwood, "United We Stand?" *The Chronicle of Higher Education* 49.24 (2003): A10.

6. See AAUP reports, "Guidelines for Good Practice: Part-Time and Non-Tenure-Track Faculty," online at http://www.aaup.org/Issues/part-time/Ptguide.htm, and "Contingent Appointments and the Academic Profession," online at http://www.aaup.org/statements/SpchState/contingent.htm. Both accessed September 12, 2004.
7. Nelson and Watt, *Academic Keywords: A Devil's Dictionary for Higher* Education (New York and London: Routledge, 1999), 210.
8. Nelson and Watt, *Academic Keywords*, 36.
9. Available online at the AAUP's website: see http://www.aaup.org/statements/Redbook/1940stat.htm.
10. Walter Broughton and William Conlogue, "What Search Committees Want." *Profession* (2001): 39–51, 44. My italics.
11. Even the average of 55 hours per week is lowered by the Department's inclusion of two-year institutions in their study (where research is only rarely required. Indeed, the average number of hours worked by professors at two-year institutions is about 49, versus 56 at research universities. See U.S. Department of Education, NCES. National Study of Postsecondary Faculty (NSOPF: 1999), Data Analysis System.
12. Kathryn Hume, *Surviving Your Academic Job Hunt: Advice for Humanities PhDs* (New York: Palgrave Macmillan), 2005.

Chapter Two

1. Cary Nelson and Stephen Watt, *Academic Keywords: A Devil's Dictionary for Higher Education* (New York and London: Routledge, 1999), 73.
2. E. D. Hirsch, *Cultural Literacy: What Every American Needs to Know* (Boston: Houghton Mifflin, 1987), 2.
3. From the department's website at http://www.virginia.edu/art/homepage/gradPhD.html.

Chapter Three

1. Xenophon, *Memorabilia*, trans. Amy L. Bonnette (Ithaca, NY: Cornell University Press, 1994), 3.12.6.

Chapter Five

1. Sean C. Grass, *The Self in the Cell: Narrating the Victorian Prisoner* (New York and London: Routledge, 2003), 7.
2. Edward Muir and Guido Ruggiero, *Microhistory and the Lost Peoples of Europe*, trans. Eren Branch (Baltimore and London: Johns Hopkins University Press, 1991), xviii.
3. William Germano, *Getting it Published: A Guide for Scholars and Anyone Else Serious about Serious Books* (Chicago and London: University of Chicago Press, 2001), 70.

4. Gregory M. Colón Semenza, *Sport, Politics, and Literature in the English Renaissance* (Newark, DE. and London: University of Delaware Press, 2003), 139.
5. Stanley Fish, *Surprised by Sin: The Reader in "Paradise Lost,"* (1967), 2nd edition (Cambridge: Harvard University Press, 1997), lxxi.
6. Eve Kosofsky Sedgwick, *Between Men: English Literature and Male Homosocial* Desire (New York: Columbia University Press, 1985), 17.
7. Stephen Jay Gould, *The Mismeasure of Man* (1981), revised ed. (New York and London: W.W. Norton and Co., 1996), 52–53.
8. Greenblatt, *Shakespearean Negotiations: The Circulation of Social Energy in Renaissance England* (Berkeley, CA: U of Berkeley P, 1989), 1.
9. Gould, *Mismeasure*, XXX.
10. Empson, *Milton's God* (Norfolk, CT: New Directions, 1961), 10–11.
11. Rorty, *Contingency, Irony, and Solidarity* (Cambridge: Cambridge UP, 1989), XXX.
12. David Kathman, "Grocers, Goldsmiths, and Drapers: Freemen and Apprentices in the Elizabethan Theater," *Shakespeare Quarterly* 55 (2004): 2.
13. William Poole, "False Play: Shakespeare and Chess," *Shakespeare Quarterly* 55 (2004): 50.
14. Norman Rabkin, *Shakespeare and the Problem of Meaning* (Chicago: U of Chicago P, 1981), 62.
15. Greenblatt, *Shakespearean Negotiations*, 65.
16. Helgerson, *Forms of Nationhood: The Elizabethan Writing of England* (Chicago: U of Chicago P, 1995), 245.
17. Coiro, " 'Fable and Old Song': *Samson Agonistes* and the Idea of a Po etic Career," *Milton Studies* 36 (1998): 147.

CHAPTER SIX

1. Rose, Mike, *Lives on the Boundary: A Moving Account of the Struggles and Achievements of America's Educationally Underprepared* (New York: Penguin, 1989), 29.
2. See D. L. McCabe and A. L. Makowski, "Resolving Allegations of Academic Dishonesty: Is There a Role for Students to Play?" *About Campus*, 6.1 (2001):17–21.
3. See Alfie Kohn, "The Dangerous Myth of Grade Inflation," *The Chronicle of Higher Education* 49.11 (2002): B7.
4. See "Harvard Raises the Bar to Curtail Grade Inflation," *The Chronicle of Higher Education*, June 7, 2002, sec. Students, p. A39.

CHAPTER EIGHT

1. Valerie Traub, "The (In)Significance of 'Lesbian' Desire in Early Modern England," in *Queering the Renaissance*, ed. Jonathan Goldberg (Durham and London: Duke University Press, 1994), 62.

CHAPTER TEN

1. Stephen Greenblatt Greenblatt to MLA members, May 28, 2002.
2. Kathryn Hume, *Surviving Your Academic Job Hunt: Advice for Humanities PhDs* (New York: Palgrave Macmillan, 2005).
3. William Germano, *Getting it Published: A Guide for Scholars and Anyone Else Serious about* Serious Books (Chicago: University of Chicago Press, 2001), 34.

CHAPTER ELEVEN

1. Walter Broughton and William Conlogue, "What Search Committees Want." *Profession* (2001): 39–51.
2. Paul Hanstedt, "Service and the Life of the Small-School Academic." *Profession* (2003): 76–84, 78.
3. Broughton and Conlogue, "What Committees Want," 44.
4. Piper Fogg, "So Many Committees, So Little Time." *Chronicle of Higher Education* 50.17 (2003): A14.
5. From The Survey of Earned Doctorates (SED), a report of "Doctorate Recipients from United States Universities: Summary Report 2002," which is sponsored by the National Science Foundation, the National Institutes of Health, the U.S. Department of Education, the National Endowment for the Humanities, the U.S. Department of Agriculture, and the National Aeronautics and Space Administration. See page 14. This useful survey is accessible online at http://www.norc.uchicago.edu/issues/sed-2002.pdf.

CHAPTER TWELVE

1. John Guillory, "Preprofessionalism: What Graduate Students Want." *ADE Bulletin* 113 (Spring 1996): 4–8.
2. David Chioni Moore, "Timing a First Entry Onto the Academic Job Market: Guidelines for Graduate Students Soon to Complete the Ph.D." *Profession* (1999): 268–74, 274.
3. Kathryn Hume, *Surviving Your Academic Job Hunt: Advice for Humanities PhDs* (New York: Palgrave Macmillan, 2005).
4. Walter Broughton and William Conlogue, "What Search Committees Want." *Profession* (2001): 39–51, 45.

AFTERWORD

1. Plato, *Republic,* trans, Desmond Lee (New York: Penguin, 1987), p. 262 519d.

APPENDIX

1. For one of many examples of this scholarly trend see Joyce Hill, "The Soldier of Christ in Old English Prose and Poetry," *Leeds Studies in English* n.s.12 (1981 for 1980 and 1981): 57–80.
2. Though I am much indebted, as I will show below and in my dissertation proper, to Johan Huizinga's *Homo Ludens: A Study of the Play-Element in Culture* because of its insightful, sociological/anthropological analysis of the origins of sport and athletics, I find in his work a possible reason for scholarship's tendency to privilege the warrior over the athlete. In his chapter "Play and War," Huizinga argues that the play-element is inherent in war, that the notion of "contest" lays at the heart of war. Inadvertently, this argument suggests that play/sport is a part of and, indeed, subsumed by war. Huizinga's conclusions were far-reaching and very influential with those scholars who succeeded him. Taking his assertions along with the fact that medieval sports were often used to train men for war (see the work of John Marshall Carter for evidence of this) may help to explain why medievalists have devoted so much time to studying war at the expense of sport in general and spiritual war instead of spiritual athletics in particular.

   I will discuss the distinction between spiritual warriors and athletes in much greater detail in the introduction of my project. For now, I will say briefly that while spiritual warriors are concerned with active pursuit of victory in Christ's name, spiritual athletes gradually condition their bodies and souls so that when temptation challenges them, they might overcome it. This focus on the body is absent from the military tradition.

Index

Boldface indicates those pages where actual samples of the specified item can be located.

CPSIA information can be obtained at www.ICGtesting.com
Printed in the USA
BVOW070842281011

274589BV00002B/3/P